COMPOSITE MATERIAL RESPONSE: CONSTITUTIVE RELATIONS AND DAMAGE MECHANISMS

Proceedings of a Workshop on 'Composite Material Response: Constitutive Relations and Damage Mechanisms', held at the Stakis Grosvenor Hotel, Glasgow, UK, July 30th and 31st 1987.

COMPOSITE MATERIAL RESPONSE:

CONSTITUTIVE RELATIONS AND DAMAGE MECHANISMS

Edited by

G. C. SIH

Institute of Fracture and Solid Mechanics,
Lehigh University, Bethlehem, Pennsylvania, USA

G. F. SMITH

Center for the Application of Mathematics,
Lehigh University, Bethlehem, Pennsylvania, USA

I. H. MARSHALL

Department of Mechanical and Production Engineering,
Paisley College of Technology, Paisley, UK

and

J. J. WU

US Army Research Office, Research Triangle Park, North Carolina, USA

ELSEVIER APPLIED SCIENCE
LONDON and NEW YORK

ELSEVIER APPLIED SCIENCE PUBLISHERS LTD
Crown House, Linton Road, Barking, Essex IG11 8JU, England

Sole Distributor in the USA and Canada
ELSEVIER SCIENCE PUBLISHING CO., INC.
52 Vanderbilt Avenue, New York, NY 10017, USA

WITH 15 TABLES AND 69 ILLUSTRATIONS

British Library Cataloguing in Publication Data

Composite material response.
1. Composite materials
I. Sih, G. C. (George C.)
620.1′18

ISBN 1-85166-228-6

Library of Congress Cataloging-in-Publication Data

Workshop on "Composite Material Response: Constitutive Relations and
Damage Mechanisms" (1987: Glasgow, Strathclyde)
Composite material response: constitutive relations and damage
mechanisms/edited by G. C. Sih ... [et al.].
p. cm.
"Proceedings of a Workshop on 'Composite Material Response:
Constitutive Relations and Damage Mechanisms,' held at the Stakis
Grosvenor Hotel, Glasgow, U.K., July 30th and 31st, 1987."
Includes index.
ISBN 1-85166-228-6
1. Composite materials—Testing—Congresses. I. Sih, G. C.
(George C.) II. Title.
TA418.9.C6W67 1987
620.1′18—dc19

Photoset in Malta by Interprint Limited
Printed in Great Britain at the University Press, Cambridge

Preface

Demand for advances in the theoretical and experimental understanding of high performance materials has provided the impetus to organize this Workshop on 'Composite Material Response: Constitutive Relations and Damage Mechanisms'. It was held at the Stakis Grosvenor Hotel in Glasgow from July 30 to 31, 1987. Experts from six different countries including Australia, Greece, Italy, Poland, United Kingdom and United States were invited to present their most recent research findings and ideas on composite material behaviour. Particular emphases were placed on the damage mechanisms associated with mechanical, thermal and/or chemical effects. Their influence on the ways with which constitutive relations are formulated is relevant in quantifying the behaviour of application-specific materials such as composites. Such a knowledge is particularly lacking.

The need for more effective use of advanced materials was emphasized in the Opening Address by Dr Fritz H. Oertel, Jr, who is Chief of Research in the US Army Research Development and Standardization Group—United Kingdom (USARDSG–UK). This Group grew out of the American, British, Canadian and Australian (ABCA) war-time alliance and sponsors research activities to the interests of the US Army and the appropriate European/Middle Eastern/African Army or Institution. To be acknowledged, in particular, is the support of this Workshop provided by the USARDSG–UK Physics/Mathematics Branch headed by Dr Julian J. Wu. It provided an opportunity for evaluating the current as well as future needs in advanced materials science and technology.

On behalf of the authors and participants, the efforts of Miss Alison Shedden connected with organizing this Workshop are greatly appreciated.

Glasgow, Scotland   G. C. SIH
                                                   G. F. SMITH
                                                   I. H. MARSHALL
                                                   J. J. WU

Contents

Preface v

1. Microstructure and Damage Dependence of Advanced
   Composite Material Behavior 1
   G. C. Sih

2. Failure Initiation in Composites from Perfectly or Partially
   Bonded Rigid Inclusions 25
   E. E. Gdoutos

3. Thermomechanical Behaviour of Composites . . . 49
   R. Jones, T. E. Tay and J. F. Williams

4. Fracture Tests for Mixed Mode Failure of Composites
   Laminates 61
   J. G. Williams

5. Constitutive Relations for Transversely Isotropic Materials . 71
   G. F. Smith and G. Bao

vii

6. Transverse Matrix Cracking in Composite Laminates . 91
   N. LAWS and G. J. DVORAK

7. Singularities in Composite Materials Applications . 101
   R. S. BARSOUM

8. Transferability of Small Specimen Data to Full-Size Structural
   Components 111
   A. CARPINTERI and P. BOCCA

9. Theoretical Modelling of Damage in Composite Laminates
   Subject to Low-Velocity Impact 133
   J. WILLIAMS, I. H. MARSHALL and W. S. CARSWELL

10. Exact Elastic Stress Analysis of Laminated Plates . . 147
    A. J. M. SPENCER

11. Recent Progress in the Mathematical Modeling of Composite
    Materials 155
    R. V. KOHN

List of Participants 179

Index 183

1

Microstructure and Damage Dependence of Advanced Composite Material Behavior

G. C. SIH

Institute of Fracture and Solid Machanics, Lehigh University, Bethlehem, Pennsylvania, USA

ABSTRACT

The macromechanical behavior of composite materials exhibits complex dependence not only on the microstructure but also in the way with which damage occurs. This is a load–time history dependent process, regardless of whether composites are being used for their many unique behaviors in terms of strength, fracture resistance, thermal properties, etc. Because the prevention of mechanical failure is almost always, an important, if not critical, requirement, damage accumulated from delamination, initiation and growth of cracks in the fiber and matrix and at the interface becomes an important consideration as the sequence of failure modes can affect the outcome. Many of the material testing procedures developed for metals may not be valid because each specimen, even though loaded uniaxially, possesses a unique behavior on account of its highly nonhomogeneous microstructure and damage pattern. Usage-specific is, in fact, a salient feature of composites that can be tailor-made to satisfy certain performance requirements.

The prediction and/or quantitative assessment of composite material behavior remain undeveloped. This is not surprising because the classical theories of mechanics and physics are not conducive for analysing inhomogeneity arising from the material microstructure and nonuniform damage. Notwithstanding such complexities, this communication will attempt to address a basic feature of composite behavior, i.e. nonuniform energy dissipation as a result of local temperature fluctuation. This is also related to change in local strain rates and strain rate history being the direct cause of microstructure inhomogeneity and damage. Load transfer characteristics as *altered by processing across fiber/matrix interfaces or bonding surfaces in laminates can alter the composite properties even when the same fibers and matrix materials are used. There is the exchange of surface and volume energy across an interface that involves the interaction of mechanical and physico-chemical effects. The fluctuation of the local volume energy density dW/dV and surface energy density dW/dA in composites would, therefore, be regarded as the quantities of interest. They are related by the rate change of volume and surface dV/dA that serves as a scaling length parameter and an index related to the degree of homogeneity or inhomogeneity of the system. Here, homogeneity includes the combined effect of material microstructure, geometry and load type. Illustrative examples will be provided and discussed briefly in connection with the technique of Electromagnetic Discharge Imaging (EDI), a possible means for evaluating the mechanical, thermal and chemical effects associated with composite material behavior.*

1. INTRODUCTION

As modern structures and vehicles are required to perform under conditions of high strength and toughness, combined with low weight, composites are becoming the material of choice. By dispersing particles or fibers of one substance in a matrix, or binder, of another, the constituent elements or microstructures can be tailor-made to meet the performance requirements. More and more of the components in military and commercial aircraft, automobiles and sports equipment are made of composites. Resistance to high temperature is another added feature of composites that is so important to the design of rocket-motor components and missile nose cones.

Composite behavior is complex because it depends on how the individual constituents are combined. The overall mechanical/thermal/electrical properties can differ widely even when composites are made from identical raw materials but processed differently. This is because processing directly affects the structure of the final product. In this context, structure refers to geometric discontinuities such as defects and voids and molecular features. Polymer matrix, for example, consists of small molecules joined chemically in chains or networks of repeating units to form huge molecules with an infinite variety of possible three-dimensional structures. The size, internal arrangement, chemical connectivity and spatial distribution of the molecules are controlled by processing. It is vital, therefore, to be able to quantify the sequence of events which culminate in composite material damage. This requires a generalized approach that can uniquely and consistently identify results at the atomic, microscopic and macroscopic scale. Material testing methods developed for metals that invoke homogeneity and isotropy are not adequate. Because a major portion of the useful life of composites involves subcritical damage in the form of fiber breaking, matrix cracking, interface debonding, delamination, etc., these mechanisms of failure cannot be ignored in predicting performance. Composite specimens will respond differently, depending on the combined interactions of load, size and geometry, in addition to the inhomogeneous nature of the microstructure.

All substances are, in fact, inhomogeneous and anisotropic when they are examined at the microscopic level. It is the sensitivity of their microstructure in responding to load that determines whether details at the lower scale level need to be analysed or not. While homogeneity can be assumed for most polycrystals deformed under normal loading conditions, the details of their microstructures become increasingly more important when the loading rate is slowed down to that of creep. (The grains in a structural steel are made sufficiently small in comparison with the specimen size so that the influence of microstructure entities can be adequately reflected via the macroparameters such as yield strength, ultimate strength, etc. The same can be done for composites if the size/time influence of the constituents are appropriately suppressed for the given loading rates under consideration.) Polycrystal metal structures are no less complicated than those in composites. Precise characterization of present-day composites that are designed to alter their global response by microstructure effects necessitates a knowledge of the way with which energy dissipates throughout the inhomogeneous structure as it is being physically damaged. This involves the continuous fluctuation of temperature as a result of the constant change of volume with surface area, dV/dA, for each composite element, an effect that has been neglected in the development of continuum mechanics theories such as elasticity, plasticity, etc. (The assumption of letting dV/dA vanish in the limit has been invoked in previous works[1-7] dealing with the failure of composite systems. This has, in fact, excluded the thermal/ mechanical interaction or the mechanism of energy dissipation. Conventional continuum mechanics theories are unable to predict the phenomenon of cooling/heating[8-11] in specimens or structures as the mechanical load is increased monotonically. Classical thermodynamics are also not valid because reversal of heat transfer leads to a sign change in entropy.)

In order to distinguish the difference between the rates at which energy is dissipated, say in a fiber and matrix, it is necessary to consider the exchange of surface and volume energy. Thermal/mechanical interactions across the interface or near a defect also give rise to temperature change and high rates of energy dissipation. To be emphasized in this work is the consumption of energy in the composite damage process that is not only time dependent but also varies from location to location. The size/time/temperature effect can obscure the observation of composite material behavior when the same event is examined at the atomic, microscopic and macroscopic scale. What appears to be cooling and dilatation on one scale level may correspond to heating and distortion on another level. Unless the fundamental hurdle of material characterization is overcome, no confidence can be placed in predicting the performance of composites. Composite damage analysis also involves the nondestructive evaluation of thermal, mechanical and chemical effects. To this end, advancements have been made on the Electromagnetic Discharge Image (EDI) technique[12,13] using high voltage electrical discharge. This method is particularly suited for analysing failure in composites. Changes in the microstructure, chemical composition, and interface condition can be identified with the mechanical behavior of composites so that a more precise distinction can be made between the intrinsic and apparent material parameters at a given scale level. A primary function of the EDI method is the evaluation of energy dissipated in various different forms.

2. THERMAL/MECHANICAL INTERACTION: FIBER, MATRIX AND INTERFACE

The thermal/mechanical/electrical behavior of composites is very much a function of the reinforcing materials and a synergy between, say, the fibers and matrix. When the fibers are stretched in the absence of the matrix, they will react individually. That is, the fibers once broken, are unable to support the load. If the fibers are embedded in a matrix that is more readily deformable, load transfer from the fibers to the matrix can prevail even when the fibers are broken. The choice of a matrix, therefore, determines not only how a composite must be fabricated. An understanding of the load transfer characteristics of the individual components in a composite is necessary before examining the behavior of the complex system.

2.1. Fiber

There are presently many materials that can be made in the form of fibers. Elements such as carbon, aluminum, silicon, etc., can be used to form compounds in which atoms are joined by strong and stable bonds. These

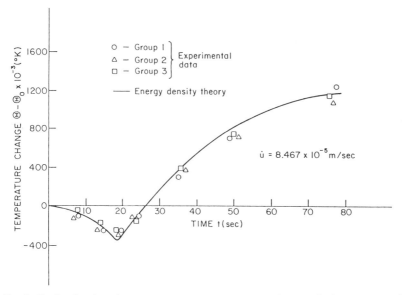

FIG. 1. Cooling/heating of 6061-T6 aluminum specimen in tension at a displacement rate of 8.467×10^{-5} m/s.

compounds will exhibit different thermal/mechanical properties when they are loaded mechanically. Figure 1 shows the temperature fluctuation, i.e. Θ as a function of time t for a 6061-T6 aluminum specimen extended with a displacement rate of 8.467×10^{-5} m/s. The specimen gauge length is approximately 57 mm and the cross-sectional area is 3.81 mm². Contrary to the ordinary notion that the material would heat up when loaded, it actually cools before returning to the ambient condition. The recovery time is approximately 26 s and is loading rate dependent. This means that energy dissipation is strictly rate dependent. It has been shown in Ref. 11 that the onset of heating can be delayed to 200 s if the loading rate is reduced by one order of magnitude. Dissipation has been known[8-11] to be small during cooling and it increases at an extremely high rate when heating starts. Therefore, mismatch of energy dissipation rates between the fiber and matrix can be undesirable as it tends to decrease the effectiveness of load transfer.

2.2. Matrix

If the same aluminum specimen discussed earlier was made of poly-carbonate and extended at the same displacement rate of 8.467×10^{-5} m/s,

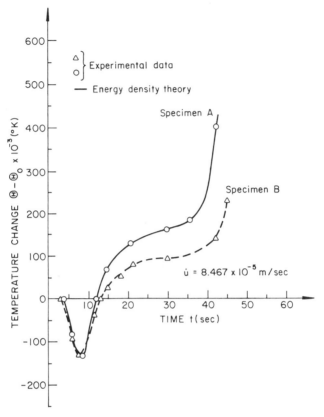

FIG. 2. Cooling/heating of polycarbonate in uniaxial extension at a displacement rate of
$8 \cdot 467 \times 10^{-5}$ m/s.

the resulting temperature as a function of time would be very different.[14]
The results are displayed in Fig. 2 in which cooling terminates at
approximately 12 s and the heating portion of the temperature curve
attained only a gradual rise before increasing rapidly. There is a delay in
energy dissipation that is in contrast to the results in Fig. 1 for the
aluminum. The time interval associated with cooling/heating is sensitive to
changes in the molecular structure. This is illustrated by the difference of the
two curves labelled specimen A and B in Fig. 2. The temperature fluctuation
data also monitor the change in the chemical composition of the hardener
molecules and the curing conditions such that the network can be altered in
predictable ways, both experimentally and theoretically. Such a capability

is made possible only by the energy density theory[8] and can be used to optimize the selection of the fiber and matrix material.

2.3. Interface

Because processing can vary the conditions at the interface, analytical modelling would not be complete without addressing both the electrochemical and mechanical bonding properties. (The assumption that displacements and stresses are continuous across the fiber and matrix is obviously unrealistic and leads to errors that depend on the type of loading.[15]) The former arises by wetting the solid surface with a fluid. Bonding of the two phases is achieved through intermolecular forces. Theoretical strengths of the order of 10^3–10^4 MPa can result within interatomic distances of a few Angstroms as shown schematically in Fig. 3. (Actual strength of an interfacial bond is lower because the classical theoretical treatment is overly idealized and is not able to include the combined effects of material, geometry and load transfer.) The Van der Waals bonds are slightly weaker with theoretical strengths of the order of 10^2–10^3 MPa and involve interatomic distances of 3–5 Å. Mechanical bonds are developed as a result of differential thermal expansion of the fiber and matrix. A compressive environment is created that results in high frictional adhesion. Compressive loading on an isolated filament in a glass fiber/epoxy composite can result in a frictional bond of 10^6 Pa. This, however, is spread over a much larger distance that is 4–5 orders of magnitude larger than the chemical bond (Fig. 3). Minute defects such as microvoids, microcracks, debonding, etc., can also prevail at the interface that further complicate the state of affairs at the fiber/matrix interface. Even though the initial interfacial energy states arising from electrochemical and

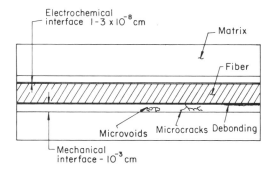

FIG. 3. Schematic of electrochemical and mechanical interface.

mechanical bonding are essential, they alone are not sufficient for characterizing the interface integrity and/or failure. The true nature of the interphase depends simultaneously on the mechanics at a two-phase boundary and rates of energy transfer that are application-specific. No generalization is possible. Each case must be analysed separately according to the specificity of the system.

The consideration of load transfer across interfaces becomes a major issue for composites in structural applications. Unless the relations between the overall bulk behavior of the composites and alleged interfacial response are known, no design guidelines can be laid down. Depending on the load type,[15] interface properties will affect the composite behavior in different ways. This is illustrated in Fig. 4(a) and (b) in which the interfacial energy

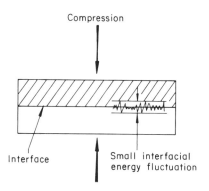

(a) Compressive load

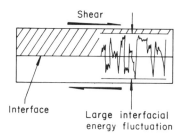

(b) Shear load

FIG. 4. Interfacial oscillation in energy due to loading: compression and shear.

oscillated much more widely in shear than that in compression. The local material elements respond at rates dictated by the properties of the two adjoining phases and the morphology in the region near the interface. Hence, the mutual synergistic effects must be regarded as unknowns and determined theoretically. The fundamental understanding on this subject has not been well developed. Apparent contradictions and/or inaccuracies can result if the interface properties are assumed to be known as *a priori*. This occurs even in the case of a single phase material where the same constitutive relation is assumed to prevail at each point of the system. (Conventional mechanics theories assume not only that constitutive relations are known but they are the same at each point in the system. This invalidates their application to problems where the local strain rates undergo large changes.) The stress and strain relations at elements near a boundary can differ widely from those inside a body.[8] They must be determined separately for each element and each time step of loading. Such an approach must be adopted for analyzing composite interface behavior.

Debonding of fiber/matrix interface should be distinguished from that of delamination between the laminae in a layered composite structure (Fig. 5). The amplitude of energy oscillation in delamination is considerably larger than that associated with debonding. Since both normal and shear stress may both be present, delamination can be best analysed by application of the energy density function dW/dV. The stationary values of dW/dV consider failure by both excessive change in shape and/or fracture as they simultaneously account for dilatation and distortion.

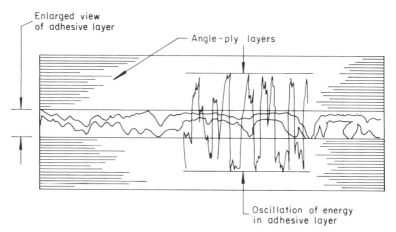

FIG. 5. Delamination of a layered composite.

3. DAMAGE OF COMPOSITES

Composite failure can be best addressed in terms of damage over an area or volume. Above all, the properties of the constituents as measured separately must be correlated with the structural features of the composite. This requires a precise knowledge of how the time dependent energy is distributed and used to deform and damage the composite system.

3.1. Failure Mechanisms

Unlike the metals where failure may be dominated by the growth of macrocracks, fiber reinforced composites fail in a cumulative fashion that may involve a combination of different modes such as fiber breaking, matrix cracking and fiber/matrix interface debonding (Fig. 6). Each of the

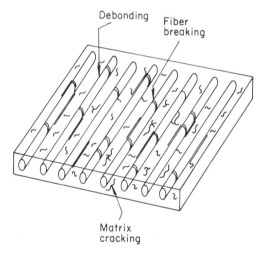

FIG. 6. Failure modes in unidirectional fiber reinforced composite.

modes may occur at a different time. Hence, the energy associated with creating new surfaces in a unit volume at a given *instance* may consist of

$$\left(\frac{dW}{dA}\right)_p = \Sigma\left[\left(\frac{dW}{dA}\right)_f + \left(\frac{dW}{dA}\right)_{f/m} + \left(\frac{dW}{dA}\right)_m\right] \quad (1)$$

in which the subscripts f, m and f/m stand, respectively, for the fiber, matrix and fiber/matrix interface. (Other forms of surface energy associated with damage may be involved such as $(dW/dA)_a$ that arises due to delamination or breaking of the adhesive in a layered composite.) Any one of the component surface energy densities, say $(dW/dA)_{f/m}$, may be zero within a certain time interval if debonding does not occur. In other words, the proportion of fiber breaking, matrix cracking and fiber/matrix debonding changes with time. What is available to create new damage surface at the next time increment is $(dW/dA)^*$. Hence, the total surface energy density is

$$\frac{dW}{dA} = \left(\frac{dW}{dA}\right)_p + \left(\frac{dW}{dA}\right)^* \tag{2}$$

The damage accumulated in a unit volume of composite element can be obtained from

$$\left(\frac{dW}{dA}\right)_i = \left(\frac{dV}{dA}\right)_i \frac{dW}{dV}, \quad i = \xi, \eta, \zeta \tag{3}$$

with dW/dV being the area under the true stress and true strain curve. The rate change of volume with surface, $(dV/dA)_i$, is made proportional to the respective slopes of the stress and strain curves. In this way, the energy state for each composite element can be uniquely determined and identified with data that are obtainable from uniaxial tests. This is accomplished by referring the components $(dW/dA)_\xi$, $(dW/dA)_\eta$ and $(dW/dA)_\zeta$ in Fig. 7 to the

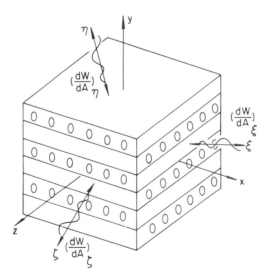

FIG. 7. Schematic of a unit volume of composite element.

damage planes such that

$$A\left(\frac{dW}{dA}\right)_\xi = B\left(\frac{dW}{dA}\right)_\eta = C\left(\frac{dW}{dA}\right)_\zeta \qquad (4)$$

The parameters A, B and C are determined by performing three different uniaxial tests along the axes of symmetry for the composite system. In this way, the constitutive relation for each composite element will be derived rather than assumed as it is now being done by the classical approach.

Thresholds for characterizing the integrity of composite systems can thus be established by considering the rate at which the volume and surface energy density approach their respective critical values. Because of eqn. (4), it suffices to consider $(dW/dA)_\xi$. Failure of composite elements is assumed when

$$\left(\frac{dW}{dA}\right)_\xi > \left(\frac{dW}{dA}\right)_c; \quad \frac{dW}{dV} \simeq \left(\frac{dW}{dV}\right)_c \quad \text{or} \quad \frac{dW}{dV} < \left(\frac{dW}{dV}\right)_c \qquad (5)$$

and/or

$$\left(\frac{dW}{dV}\right) > \left(\frac{dW}{dV}\right)_c; \quad \left(\frac{dW}{dA}\right)_\xi \simeq \left(\frac{dW}{dA}\right)_c \quad \text{or} \quad \left(\frac{dW}{dA}\right)_\xi < \left(\frac{dW}{dA}\right)_c \qquad (6)$$

Depending on the rate of load transfer in the composite, the exchange of surface and volume energy density can vary over a wide range as indicated in eqns (5) and (6) so that failure mechanisms involving fiber breaking, matrix cracking, interface debonding, etc., can be quantified in terms of temperature change, energy dissipation and the rate change of volume with surface.

Related to $(dW/dA)_p$ and $(dW/dA)^*$ are the quantities $(dW/dV)_p$ and $(dW/dV)^*$ which are, respectively, the dissipated and available volume energy as defined in Fig. 8 such that

$$\frac{dW}{dV} = \left(\frac{dW}{dV}\right)_p + \left(\frac{dW}{dV}\right)^* \qquad (7)$$

Note that $(dW/dV)_p$ or $\mathscr{D}$ is the area oypq and $(dW/dV)^*$ or $\mathscr{A}$ the area pgq. The path of unloading is determined analytically in the energy density theory. It is of interest to point out that energy dissipation in composite specimens has been measured experimentally.[16] (One of the experimental curves for $\mathscr{D}$ in Ref. 16 became negative which is a physical impossibility and obviously caused by error in measurement.) It, of course, must be

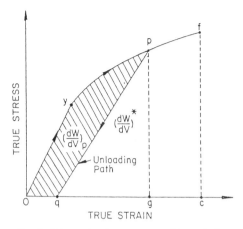

FIG. 8. Schematic of true stress and true strain.

positive definite and increase monotonically with time, i.e.

$$\mathscr{D} \geqslant 0; \quad \frac{\mathrm{d}\mathscr{D}}{\mathrm{d}t} \geqslant 0 \tag{8}$$

The method of Electromagnetic Discharge Imaging (EDI) can also be used. Once $\mathscr{D}$ is known, the temperature Θ can be obtained from

$$\frac{\Delta\Theta}{\Theta} = -\lambda \frac{\Delta V}{\Delta A} \frac{\Delta\varepsilon}{\Delta\mathscr{D}/\Delta\varepsilon} \tag{9}$$

in which $\Delta\varepsilon$ is the increment of strain. It can be referred to the plane of homogeneity (this is the plane defined in Fig. 7 such that $(\mathrm{d}W/\mathrm{d}A)_i$ for $i = \xi, \eta, \zeta$ are related as shown in eqn. (4)) with $\Delta\varepsilon_\xi$ being the equivalent uniaxial incremental strain for a multiaxial stress or strain state.

Equation (9) has been used successfully for determining many of the previously unexplained thermal/mechanical interaction effects[9-11] in isotropic and homogeneous materials. It applies equally well to composites. The quantity Θ is not equal to the temperature T defined by classical thermodynamics. It is equal to T only when all energies are dissipated in the form of heat Q. In general, $\mathscr{D}$ is not equal to Q. The strain rate of energy dissipation $\Delta\mathscr{D}/\Delta\varepsilon$ can be associated with the latent heat in classical thermodynamics for determining phase transformation. Again, only when $\mathscr{D} \to Q$ will the following relation hold:

$$\frac{\Delta\mathscr{D}}{\Delta\varepsilon} \to \frac{\Delta Q}{\Delta v} = T\frac{\Delta\mathscr{S}}{\Delta v} \tag{10}$$

Here, ΔQ is the heat required per unit mass of substance in changing the thermodynamic state from the temperature T to $T + \Delta T$ and Δv is the change of specific volume. The corresponding change in entropy is $\Delta \mathscr{S}$. Phase transformation is a rate dependent process, an effect not considered in classical thermodynamics.

3.2. Characterization

A unique characterization of microfailure mechanisms in composites can be made by specifying the quantities $\Delta V/\Delta A$, $\Delta \Theta$ and $\Delta \mathscr{D}$, all of which can be measured individually and independently. Their combination can uniquely determine any changes in the composite microstructure. There are no difficulties at present in measuring $\Delta V/\Delta A$ via a knowledge of the displacement field and local temperature changes within $\pm 10^{-3}$ to $\pm 10^{-4}$ K in microelements having linear dimensions of $10^{-2} - 10^{-3}$ cm. The response time of the electronic instruments must be adjusted in accordance with the loading rate that governs the way with which the material microstructure reacts. A major challenge is the detection of $\Delta \mathscr{D}$. Being developed at the Lehigh Institute of Fracture and Solid Mechanics is the electromagnetic discharge imaging (EDI) technique.[17] It involves accelerating the electrons surrounding the object or specimen by an external field such that the air molecules are ionized to enhance an exponential growth in the number of electrons and positive ions causing an avalanche.

To fix ideas, consider the composite system in Fig. 9 consisting of three different types of microfailure, namely, fiber breaking, matrix cracking and

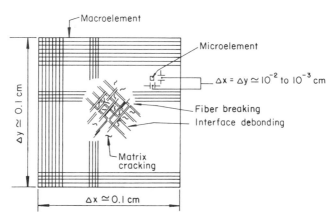

FIG. 9. Hypothetical microfailure in fiber reinforced composite.

fiber/matrix debonding in addition to the texture of fiber and matrix. A sufficient number of fibers is assumed to contain in the macroelement of, say $\Delta x = \Delta y \simeq 10^{-1}$ cm, such that the microfailure response can be reflected by the time-dependent macroscopic parameters. The state of affairs at the microscopic level is much more complex because microheterogeneities are now of the same size order as the microelements, say $\Delta x = \Delta y \simeq 10^{-2} - 10^{-3}$ cm. Table 1 outlines the various combinations. The variations of

TABLE 1

A hypothetical characterization of microelements for a fiber reinforced composite at time t_p

Microelement type	$(\Delta V/\Delta A)_\xi$	$\Delta\Theta$	$\Delta\mathscr{D}$
Matrix	Slightly above	Slightly below	Slightly below
Fiber	Above	Below	Above
Interface	Slightly below	Slightly above	Near average
Fiber/matrix	Near average	Near average	Slightly above

$\Delta V/\Delta A$, $\Delta\Theta$ and $\Delta\mathscr{D}$ for the four different types of microelements can be compared with the average values obtained by considering all the microelements in the macroelement. The loading at the instance t_p may be such that $\Delta V/\Delta A$ and $\Delta\mathscr{D}$ for the undamaged composite element are only slightly above the average while $\Delta\Theta$ is slightly below. This is illustrated in Figs. 10(a)–(c). For the element damaged by fiber breaking and matrix cracking, the deviations of $\Delta V/\Delta A$, $\Delta\Theta$ and $\Delta\mathscr{D}$ from the average may retain the same trend but become more pronounced as shown in Figs. 11(a)–(c). In the case of interface debonding in Figs. 12(a)–(c), $\Delta V/\Delta A$ may be lower and $\Delta\Theta$ higher than the average. This can be accompanied by

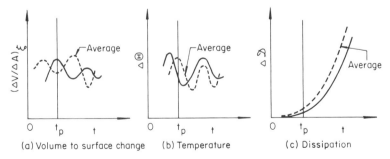

(a) Volume to surface change   (b) Temperature   (c) Dissipation

FIG. 10. Time response of undamaged composite microelement.

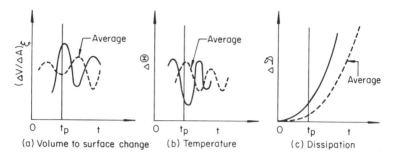

FIG. 11. Time response of composite microelement damaged by fiber breaking and matrix cracking.

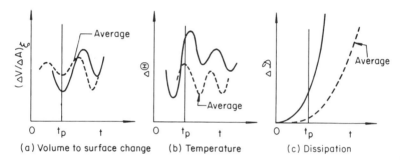

FIG. 12. Time response of composite microelement with interface debonding.

a higher than average dissipation. Each microelement will exhibit a uniquely different microstress and microstrain response even for the same microstructure detail because the load distribution is, in general, not uniform.

Microstructure change can also take place by phase transformation. This rate process depends on the strain rate dissipation energy density $\Delta\mathcal{D}/\Delta\varepsilon$ in eqn. (10) at a given temperature Θ. Initiation of phase transformation can be identified with the reversal of curvature of the $\mathscr{H}$-curve. The corresponding time t_q is shown in Fig. 13(a). The values of $(\Delta\mathcal{D}/\Delta\varepsilon)_q$ and Θ_q at t_q can thus be obtained from the curves in Fig. 13(b) and (c), respectively. In this way, the distribution and change of the different phases at a given location can be assessed analytically and experimentally.

3.3. Scale of Observation

The transient character of size/time/temperature interaction has eluded those working in composite materials. Failure modes observed at one scale

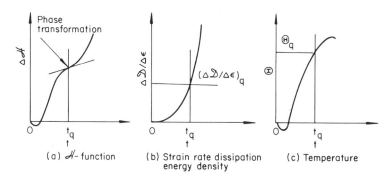

FIG. 13. Assessment of local phase transformation from the $\mathscr{H}$-curve.

level may differ from that seen at another level and they are further complicated by changes in loading rates. In general, dilatation/distortion and cooling/heating tend to flip-flop as the scale level of observation is altered; depending, of course, also on the time response. The fundamentals of this alternating mechanism are discussed in Ref. 18 and will only be mentioned briefly in relation to an element dominated by macrodilatation and another by macrodistortion as shown, respectively, by Figs. 14(a) and (b). (The combination of size/time/temperature data are selected arbitrarily for this discussion. Actual values for uniaxial tensile and compressive metal specimens have been obtained and can be found in Ref. 19.) These two elements are necessarily located at different locations of the same system. To be emphasized in Fig. 14(a) is that macrocooling and macrodilatation for response times of 1–10 s correspond to microheating and microdistortion when the same event is viewed within the time interval of 10^{-2}–10^{-1} s. (Macro refers to dimensions of 10^{-3}–10^{-2} cm; micro to 10^{-4}–10^{-5} cm; and atomic to 10^{-6}–10^{-8} cm. The size/time/temperature scale can shift depending on the loading or local strain rate.) Disturbances at the atomic scale refer to even smaller response time and temperature fluctuation. The situation for the element in Fig. 14(b) is opposite to that in Fig. 14(a). What was macrocooling now becomes macroheating. The same applies to the micro- and atomic-scale together with the corresponding time response and temperature. Scaling of size/time/temperature must be assessed quantitatively such that seemingly different behavior of the same failure process when viewed at the atomic, microscopic and macroscopic level can be related uniquely by the three variables mentioned earlier. Microstructure change must be addressed in its entirety with reference to the specimen response. Whether its influence could be adequately reflected by the macroscopic variables or not depends on the specific application.

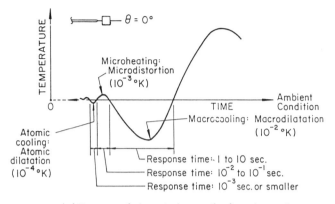

(a) Response of element along path of crack growth

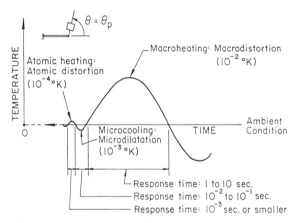

(b) Response of element along direction of maximum shape change

Fig. 14. Size/time/temperature response of element at different scale level. (a) dominated by macrodilatation and (b) dominated by macrodistortion.

4. ELECTROMAGNETIC DISCHARGE IMAGING: NONDESTRUCTIVE EVALUATION

Because of the complex nature of composite failure behavior, a variety of analytical models have emerged in an attempt to explain the different failure modes. Since no unified treatment has come forth, damage inspection procedures have relied solely on the development of *hardwares* with the assumption that the data related to debonding, fiber and matrix degradation, etc., will eventually be understood and analysed by theories. Among the techniques used are ultrasonics, radiography, acoustic emission, thermography, eddy current and a host of others. Such a procedure has been loosely referred to as Non-Destructive Testing (NDT). What is urgently needed is, of course, a consistent way of assessing composite damage so that the seemingly different failure modes can be weighed on a common basis. There is no other way for qualifying the integrity of composite materials and structures.

Damage being a time-rate dependent process is more widely spread throughout the composite simply because the load reacts more sensitively with material inhomogeneity. The idea of NDT is to correlate the difference between the emitted and received signal with the size and location of damage. Those energies that are already present within the system may also be released during the damage process to interrupt the NDT signal. The unique identification of energy dissipation rate with failure mode is a prerequisite for the quantitative assessment of damage, i.e. the quantity $d\mathscr{D}/dt$ or those in eqn. (1) involving $d(dW/dA)_f/dt$, $d(dW/dA)_m/dt$, etc. The Electromagnetic Discharge Image (EDI) technique[12,13] commonly known as Kirlian photography, is particularly suited for incorporating the energy density theory.[8] (This technique has been used extensively in the USSR in NDT and is rarely publicized. Preliminary efforts made by the US Navy and Lawrence Livermore Laboratory lacked the theoretical support in NDT application.) Utilized in the process is a high-voltage electrical discharge field. Electrons with the specimen are accelerated and multiply exponentially. Streamers are thus created that move at speeds of $10^7 - 10^8$ cm/s. In air, at high field strength, the streamers are blue while a reddish-purple glow appears in low electric fields. The intensity, color and pattern of the discharge image can be recorded photographically or digitally and identified with the mechanical, thermal, chemical and electrical disturbances in composites as they are damaged. Mechanical discontinuities in the range of 10^{-1}–10^{-6} mm can be detected. Such a wide range of detection capability is unmatched by the other conventional methods. The flow chart in Fig. 15 shows how the available (or dissipated) energy is monitored by EDI. A laboratory EDI model has been designed by the Institute of Fracture and Solid Mechanics at Lehigh University. The combined thermal, mechanical, chemical and electrical effects are recorded everywhere on a specimen as indicated in Fig. 16(a). Spectral analysis of the pixel can then be made to obtain the optical intensity as a function of radiation wavelength (Fig. 16(b)). For each radiation wavelength λ_j, the optical intensity I_j can be isolated and identified with a given energy level U_j. This

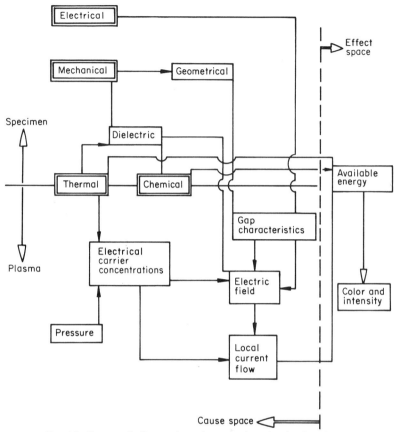

FIG. 15. Cause and effect on intensity and color of emitted radiation.

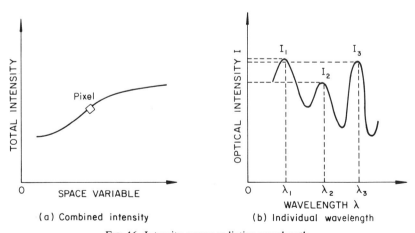

(a) Combined intensity

(b) Individual wavelength

FIG. 16. Intensity versus radiation wavelength.

is accomplished by means of a plasma spectral analysis, the results of which are shown schematically in Fig. 17(a)–(c). The results can be further processed and displayed by contour plots in two or three dimensions as illustrated in Fig. 18 by which the individual thermal, mechanical, chemical

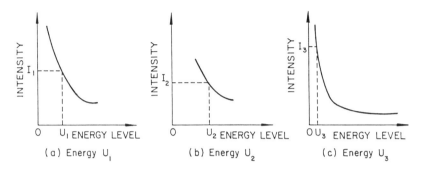

FIG. 17. Variations of intensity with energy level.

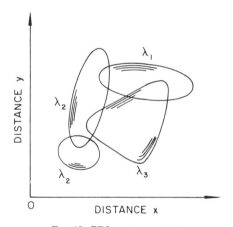

FIG. 18. EDI contour output.

or electrical effect can be sorted out. The color, intensity and distribution of the contours provide information on the location and magnitude of the stationary values of the energy density whose relation with material damage has been discussed earlier. These contours will change as a function of time, if load is present and hence the influence of time-rate dependency on damage can also be evaluated. To be more specific, oscillation of the electrochemical bond energy may be displayed by contour plots in a two-dimensional space. Refer to Fig. 19 in which contours with high and low energy densities represent fluctuations in the electrochemical forces. They can be measured by the EDI technique and calculated by the energy density theory.[8] As the scale of investigation is enlarged near the

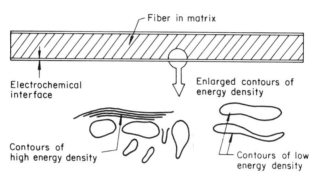

FIG. 19. Schematic of electrochemical interface energy density contours detected by EDI.

fiber/matrix interface, stationary values of the mechanical and thermal energy density can be obtained in the same way. The *same* theoretical and experimental method must be applicable to cover the range of energy dissipation rates that occur in composite material damage.

REFERENCES

1. SIH, G. C., HILTON, P. D., BADALIANCE, R., SHENBERGER, P. S. and VILLARREAL, G., *Fracture Mechanics for Fibrous Composites*, American Society for Testing and Materials, STP 521, 1973, pp. 98–132.
2. SIH, G. C., CHEN, E. P. and HUANG, S. L., Fracture mechanics of plastic-fiber composites, *Journal of Engineering Fracture Mechanics*, **6**(1974), 343–59.
3. SIH, G. C., CHEN, E. P., HUANG, S. L. and MCQUILLEN, E. J., Material characterization on the fracture of filament-reinforced composites, *Journal of Composite Materials*, **9**(1975), 167–86.
4. SIH, G. C. and TAMUZS, V. P., eds, *Fracture Mechanics of Composite Materials*, The Hague, Martinus Nijhoff, 1979.
5. SIH, G. C. and TAMUZS, V. P., eds, *Fracture of Composite Materials*, The Hague, Martinus Nijhoff, 1982.

6. SIH, G. C. and CHEN, E. P., *Cracks in Composite Materials*, The Hague, Martinus Nijhoff, 1982.
7. SIH, G. C. and SKUDRA, A. M., eds, *Failure Mechanics of Composites*, Amsterdam, North-Holland, 1985.
8. SIH, G.C., Mechanics and physics of energy density and rate of change of volume with surface, *Journal of Theoretical and Applied Fracture Mechanics*, **4**(1985), 157–73.
9. SIH, G. C. and TZOU, D. Y., Heating preceded by cooling ahead of crack: macrodamage free zone, *Journal of Theoretical and Applied Fracture Mechanics*, **6** (2) (1986), 103–11.
10. SIH, G. C. and TZOU, D. Y., Irreversibility and damage of SAFC-4OR steel specimen in uniaxial tension, *Journal of Theoretical and Applied Fracture Mechanics*, **7** (1) (1987) 23–30.
11. SIH, G. C., LIEU, F. L. and CHAO, C. K., Thermal/mechanical damage of 6061-T6 aluminum tensile specimen, *Journal of Theoretical and Applied Fracture Mechanics*, **7** (2) (1987), 67–78.
12. SIH, G. C. and MICHOPOULOS, J. G., Nondestructive detection of damage in aluminum: electromagnetic discharge imaging, *Journal of Theoretical and Applied Fracture Mechanics*, **5** (1) (1986), 23–30.
13. MICHOPOULOS, J. G. and SIH, G. C., Electromagnetic discharge imaging technique as a means of nondestructive evaluation of material imperfections, Institute of Fracture and Solid Mechanics Technical Report IFSM-85-132, 1985.
14. SIH, G. C., Cooling/heating of polycarbonate tensile specimens, Technical Report, Texas, Dow Chemical USA, February 1987.
15. SIH, G. C., Influence of interface modelling on composite failure, *Continuum Models of Discrete Systems*, University of Waterloo Press, Waterloo, Ontario, Canada, 1979, pp. 837–64.
16. MAST, P. W., BEAUTIEN, L. A., CLIFFORD, M., MULVILLE, D. R., SUTTON, S. A., THOMAS, R. W., TIROSH, J. and WOLOCK, J., A semi-automated in-plane loader for material testing, *Journal of Experimental Mechanics*, **23** (2) (1983), 236–41.
17. MICHOPOULOS, J. G. and SIH, G. C., Experimental evaluation of energy dissipation by electromagnetic discharge imaging, Institute of Fracture and Solid Mechanics Technical Report (in preparation).
18. SIH, G. C., *Thermal/Mechanical Interaction Associated with the Micromechanisms of Material Behavior*, Institute of Fracture and Solid Mechanics Monograph, Library of Congress No. 87-080715, February 1987.
19. SIH, G. C. and CHAO, C. K., Scaling of size/time/temperature associated with damage of uniaxial tensile specimens, Institute of Fracture and Solid Mechanics Technical Report (in preparation).

2

Failure Initiation in Composites from Perfectly or Partially Bonded Rigid Inclusions

E. E. GDOUTOS

School of Engineering, Democritus University of Thrace,
Xanthi, Greece

ABSTRACT

An analysis of failure of composites originating from rigid inclusions is undertaken. The following cases are considered: (i) inclusions with cuspidal points embedded in an elastic matrix and (ii) inclusions partially bonded to the matrix. In the first case failure initiates from the cuspidal points, while in the second case from the tips of the interfacial crack which coincides with the unbonded part of the inclusion due to the existing stress singularity at these points. A closed form solution for the problem of partially bonded inclusions is obtained by using the conformal mapping technique of the complex variable theory of elasticity and reducing it to a Hilbert problem. Results are obtained for special inclusion shapes including the fiber, the hypocycloidal, the astroidal, the square and the triangular inclusion. The strain energy density theory is used to study the failure of the composite and the critical load and initial fracture angles are determined.

1. INTRODUCTION

Composites made up of a soft matrix and stiff particles are analysed regarding their mechanical behavior along two distinct approaches: the macro-approach and the micro-approach. The macro-approach considers the composite as a homogeneous system which has the overall material properties of the macrostructure and applies the principles of continuum mechanics of homogeneous (isotropic or anisotropic) materials. The micro-approach takes into account the detailed geometrical and physical properties of the constituent materials and models the real system by a simpler one. The approach chosen in the analysis of composites depends on the nature of the problem and the objective of the study.

In the analysis of composites within the framework of the micro-approach certain idealizations referred to the geometry and the properties of the constituent materials are usually made. Thus, the filler particles are simulated by bodies with simple geometrical shape, like spheroidal, ellipsoidal or cylindrical. Furthermore, when their modulus of elasticity is much higher than the modulus of elasticity of the matrix the particles are considered stiff.

In many composites the reinforcing constituents are of irregular shape with sharp angles, like the various inorganic fillers, the metal or boron filaments, the aggregate or sand particles in concrete. In such cases in the sharp angle corners, high stress concentrations develop and therefore they are nuclei for the generation of cracks and slip bands leading to failure. A commonly observed failure mode of multiphase materials is the debonding of the different phases due to manufacturing and/or loading conditions, thus forming interfacial cracks. This is usually encountered in concrete where cracks are formed along the boundaries of the aggregates which are embedded in the mortar.

In the present communication in an attempt to model the failure behavior of composites originating from rigid inclusions within the framework of micro-approach the following two problems are studied:

(i) Rigid inclusions with cuspidal points along their boundaries embedded in an elastic matrix. In such cases high stress concentrations are developed in the vicinity of the cuspidal points which constitute nuclei for failure initiation.

(ii) Rigid inclusions of general shape partially bonded to an elastic matrix. The unbonded part of the inclusion constitutes an interfacial crack from which failure of the composite starts.

In both cases the stress field in the composite is first analysed and it is then coupled with an appropriate failure criterion. Due to the importance of the problem elasticity solutions of a number of geometrical configurations of inclusions perfectly or partially bonded to a matrix have appeared in the literature. Panasyuk et al.[1] studied the case of a rigid inclusion with cuspidal points on its boundary embedded in an elastic matrix. They found that the stress field in the vicinity of the cuspidal points presents an inverse square root singularity and it is expressed in terms of two stress concentration factors. The problem of a rigid inclusion partially bonded to an elastic matrix has been studied by various researchers.[2-10] Circular, elliptical and general shaped inclusions have been considered and attention has been paid to the neighborhood of the ends of the unbonded part of the inclusion.

In this work the failure of composite plates containing a perfectly or partially bonded rigid inclusion in an elastic matrix is analysed. A general solution for the stress and displacement field is obtained for a curvilinear inclusion with an interfacial crack using the method of conformal mapping in conjunction with the analytic continuation of the complex potentials. The problem is reduced to a Hilbert problem for one of the complex potentials and general formulae for the determination of the various unknown coefficients of the solution are given. Special cases include the square and the triangular inclusions. The results of the stress analysis are combined with the strain energy density failure criterion.[11,12] The critical load for fracture initiation and the more vulnerable failure sites are determined. The failure characteristics of the composite as influenced by the type and geometrical configuration of the inclusion, the material properties and the type of loading are disclosed.

2. RIGID INCLUSIONS WITH CUSPIDAL POINTS

2.1. The Stress Field

Consider a rigid inclusion with a cuspidal corner O perfectly bonded to an infinite isotropic elastic plate which is subjected to a system of stresses at infinity (Fig. 1). A reference frame of Cartesian coordinates is attached to the point O with the x-axis along the tangent of the boundary of the inclusion at O. The local stress field in the vicinity of the point O is given by:[1]

$$\sigma_r = \frac{1}{4\sqrt{2r}}\left[k_1\left[5\cos\frac{\theta}{2}+(2\kappa+1)\cos\frac{3\theta}{2}\right]-k_2\left[5\sin\frac{\theta}{2}+(2\kappa-1)\sin\frac{3\theta}{2}\right]\right]$$

$$\sigma_\theta = \frac{1}{4\sqrt{2r}}\left[k_1\left[3\cos\frac{\theta}{2}-(2\kappa+1)\cos\frac{3\theta}{2}\right]-k_2\left[3\sin\frac{\theta}{2}-(2\kappa-1)\sin\frac{3\theta}{2}\right]\right]$$

$$\tau_{r\theta} = \frac{1}{4\sqrt{2r}}\left[k_1\left[\sin\frac{\theta}{2}-(2\kappa+1)\sin\frac{3\theta}{2}\right]+k_2\left[\cos\frac{\theta}{2}-(2\kappa-1)\cos\frac{3\theta}{2}\right]\right] \quad (1)$$

where the coefficients k_1 and k_2 are independent of the coordinates r, θ and

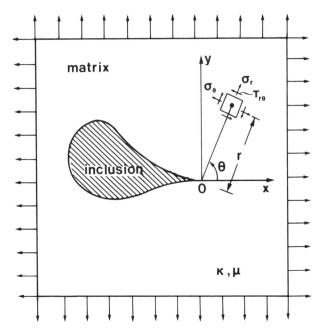

FIG. 1. Geometry of a general shaped inclusion with a singular corner O embedded in a matrix.

depend only on the loading conditions, the material of the plate and the geometrical shape of the inclusion at the cuspidal corner. In the above relations, $\kappa = (3-v)/(1+v)$ or $\kappa = 3-4v$ for plane stress or plane strain conditions respectively, with v representing the Poisson's ratio of the material of the plate. Note the inverse square root singularity of the stress field near the point O.

Failure of the composite plate initiates from the cuspidal point O due to the high intensification of the stress field around this point. For the determination of the critical failure loads and the fracture path use is made of the minimum strain energy density criterion.

2.2. The Minimum Strain Energy Density Criterion

Failure initiation from the cuspidal point of the inclusion is described by the minimum strain energy density criterion proposed by Sih[11,12] and used for the solution of a host of mixed-mode crack problems by Gdoutos.[13]

The fundamental quantity for unstable fracture is the strain energy density factor S defined by

$$S = r\frac{dW}{dV} \tag{2}$$

where dW/dV is the strain energy density per unit volume and r is the distance from the cuspidal point. For the plane elastic problem dW/dV is expressed by

$$\frac{dW}{dV} = \frac{1}{4\mu}\left[\frac{(\kappa+1)(\sigma_r+\sigma_\theta)^2}{4} - 2(\sigma_r\sigma_\theta - \tau_{r\theta}^2)\right] \tag{3}$$

where μ represents the shear modulus.

It is assumed that the fracture path from the cuspidal point of the inclusion follows the direction of minimum strain energy density factor, defined by

$$\frac{\partial S}{\partial \theta} = 0 \qquad \frac{\partial^2 S}{\partial \theta^2} > 0 \tag{4}$$

Unstable fracture occurs when the minimum value S_{min} of S becomes equal to a critical value S_{cr} which is a material constant, that is

$$S_{min} = S_{cr} \tag{5}$$

Relations (4) define the fracture path originating from the cuspidal point, while eq. (5) gives the critical load.

From eqns (1)–(3) is obtained

$$S = a_{11}k_1^2 + 2a_{12}k_1k_2 + a_{22}k_2^2 \tag{6}$$

where the coefficients $a_{ij}\,(i,j=1,2)$ are given by:

$$16\mu a_{11} = 2(\kappa-1)\cos^2\frac{\theta}{2} + \kappa^2 + (2\kappa+1)\cos^2\theta$$
$$16\mu a_{12} = -[(\kappa-1) + 2\kappa\cos\theta]\sin\theta \tag{7}$$
$$16\mu a_{22} = 2(\kappa-1)\sin^2\frac{\theta}{2} + \kappa^2 - (2\kappa-1)\cos^2\theta$$

The minimum strain energy density criterion is used for the determination of fracture of a composite plate with a linear, an astroidal and a hypocycloidal inclusion.

2.3. The Linear Inclusion

A rigid rectilinear inclusion of length $2l$ is embedded in an elastic plate which is subjected to a uniform uniaxial stress σ making an angle β with the axis of the inclusion. The stress concentration factors k_1 and k_2 are given by [1]

$$k_1 = \frac{\sigma\sqrt{l}}{2\kappa}\left(\frac{\kappa-1}{2} + \cos 2\beta\right)$$

$$k_2 = -\frac{\sigma\sqrt{l}}{2\kappa}\sin 2\beta$$

(8)

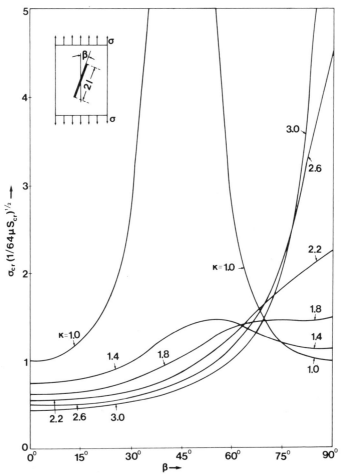

FIG. 2. Variation of the normalized critical stress of fracture $\sigma_{cr}/(l/64\mu S_{cr})^{1/2}$ for the case of a fiber inclusion versus its orientation angle β according to the minimum strain energy density criterion.

From eqns (4) – (7) the fracture path and the critical stress σ_{cr} for unstable fracture of the composite plate are obtained. Figure 2 presents the variation of the normalized critical stress σ_{cr} versus the inclination angle of the inclusion β for various values of the material constant κ. Note that for $\kappa = 1$, σ_{cr} is symmetrical with respect to $\beta = 45°$ for which σ_{cr} becomes infinite. For $\kappa = 1.4$ and 1.8, σ_{cr} becomes maximum in the interval $0 < \beta < 90$, while for the remaining values of κ it takes its maximum value for $\beta = 90°$. The variation of the initial fracture angle θ_0 is shown in Fig. 3.

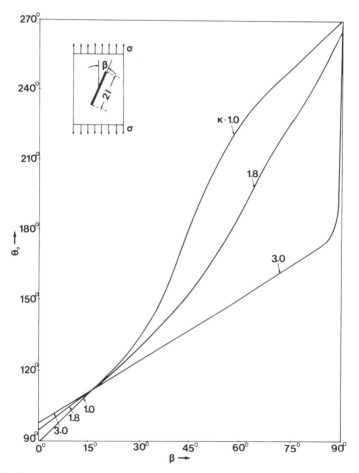

FIG. 3. Variation of the fracture angle θ_0 versus the angle β of the orientation of the fiber according to the minimum strain energy density criterion.

2.4. The Astroidal Inclusion

An astroidal inclusion is circumscribed by a circle of radius a and it is referred to a reference frame Oxy with O coinciding with the center of the circle and the singular corner $j=0$ lying on the x-axis. The equation of the inclusion with respect to the frame Oxy is given by

$$z = \frac{3a}{4}\left(\zeta + \frac{1}{3}\zeta^{-3}\right) \tag{9}$$

with $z = re^{i\phi}$ and $\zeta = e^{i\theta}$.

If β is the angle that the applied stress σ subtends with the x-axis, then the k_1 and k_2 stress concentration factors are given by[1]

$$k_1^{(j)} = \frac{\sqrt{3a}}{4\kappa}\sigma\left[\frac{\kappa-1}{2} + \frac{3\kappa}{3\kappa+1}\cos(\pi j - 2\beta)\right]$$

$$\tag{10}$$

$$k_2^{(j)} = \frac{\sqrt{3a}}{4}\sigma\frac{3}{3\kappa-1}\sin(\pi j - 2\beta)$$

with $j = 0, 1, 2, 3$ for the four cuspidal points of the inclusion.

Because of the existing symmetry only the critical fracture stresses $\sigma_{cr}^{(j)}$ from the points $j=0$ and $j=1$ are determined. Fracture of the composite plate starts from the more vulnerable cuspidal point which has the smaller critical stress. Figure 4 presents the variation of the dimensionless quantity $\sigma_{cr}(3a/256\mu S_{cr})^{1/2}$ versus angle β. Regions where fracture starts from either of the cuspidal points $j=0$ or $j=1$ are separated by a dotted line.

2.5. The Hypocycloidal Inclusion

The hypocycloidal inclusion circumscribed by a circle of radius a and referred to a reference frame Oxy with O coinciding with the center of the circle and the Ox axis being the axis of symmetry of the inclusion is described by the following equation.

$$z = \frac{2a}{3}\left(\zeta + \frac{1}{2}\zeta^{-2}\right) \tag{11}$$

with $z = re^{i\phi}$ and $\zeta = e^{i\theta}$.

If β is the angle that the applied stress σ subtends with the x-axis, then the

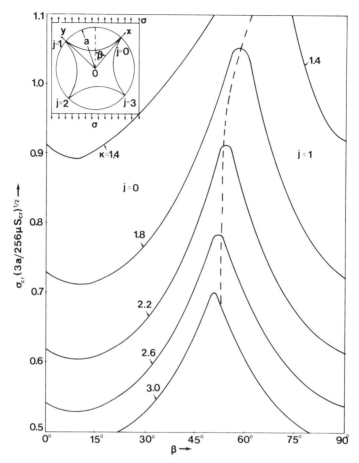

FIG. 4. Variation of the normalized critical stress of fracture $\sigma_{cr}(3a/256\mu S_{cr})^{1/2}$ versus angle β for the case of an astroidal inclusion according to the minimum strain energy density criterion. Regions where fracture starts from either of the corners $j=0$ or $j=1$ are indicated.

k_1 and k_2 stress concentration factors are given by [1]

$$k_1^{(j)} = \frac{\sqrt{2a}}{3\kappa}\sigma\left[\frac{\kappa-1}{2} + \cos\left(\frac{4\pi j}{3} - 2\beta\right)\right]$$

$$k_2^{(j)} = \frac{\sqrt{2a}}{3\kappa}\sigma\sin\left(\frac{4\pi j}{3} - 2\beta\right)$$

(12)

with $j=0$, 1, 2 for the three cuspidal points of the inclusion.

As in the previous cases the critical stress σ_{cr} and the fracture angle θ_0 from the more vulnerable cuspidal point of the inclusion are determined. The variation of the quantity $\sigma_{cr}(a/72\mu S_{cr})^{1/2}$ versus angle β for fracture initiation from the cuspidal points $j=0$ and $j=2$ is shown in Fig. 5.

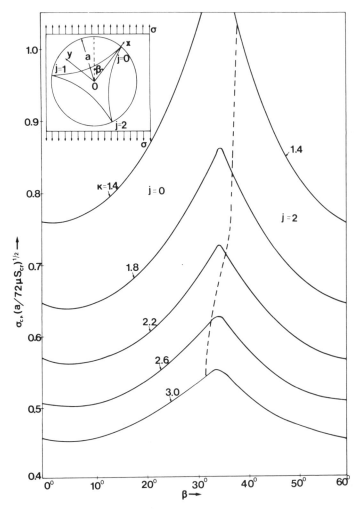

FIG. 5. Variation of the normalized critical stress of fracture $\sigma_{cr}(a/72\mu S_{cr})^{1/2}$ versus the orientation angle of the inclusion β for the case of a hypocycloidal inclusion. Regions where fracture starts from either of the corners $j=0$ or $j=2$ are indicated.

3. PARTIALLY BONDED RIGID INCLUSIONS

3.1. Statement of the Problem and Hilbert Formulation

Consider a rigid curvilinear inclusion perfectly bonded to an elastic infinite matrix except from a part A_S of its boundary forming an interfacial crack (Fig. 6). Denote by A_D the bonded part of the inclusion boundary, and by κ and μ the elastic properties of the matrix. A system of uniformly distributed principal stresses T_∞ and N_∞ is applied at infinity where the direction of T_∞ makes an angle φ with the x-axis.

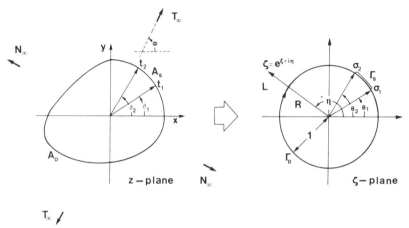

FIG. 6. Geometry of a rigid curvilinear inclusion partially bonded to an elastic matrix and its conformal mapping on to the unit circle.

The matrix occupying the z-plane is mapped in the infinite region Σ of the ζ-plane bounded by the unit circle Γ by means of the function $z = m(\zeta)$ which has the form

$$z = R\left(\zeta + \frac{b_1}{\zeta} + \frac{b_2}{\zeta^2} + \cdots + \frac{b_\rho}{\zeta^\rho}\right) \tag{13}$$

where R is a real and $b_1, b_2, \ldots, b_\rho$ are generally complex constants. The values of these constants are determined so that the contour A in the z-plane corresponds to the circumference Γ of the unit circle in the ζ-plane. The positive sense of describing the contour A is chosen to be clockwise so that the region S remains on the left when moving in the positive direction. Putting

$$\zeta = e^\xi(\cos\eta + i\sin\eta) \tag{14}$$

circles of radii $|\zeta| = e^{\xi_0}$ and straight lines $\eta = \eta_0$ in the ζ-plane define an orthogonal curvilinear coordinate system (ξ, η) in the z-plane. Denote by $t_1 = r_1 e^{i\vartheta_1}$ and $t_2 = r_2 e^{i\vartheta_2}$ the tips of the interfacial crack A_s which are mapped on the unit circle Γ at points $\sigma_1 = e^{i\theta_1}$ and $\sigma_2 = e^{i\theta_2}$. Assuming that the crack lips are stress-free the boundary conditions of the problem take the form

$$u(\sigma) + iv(\sigma) = i\varepsilon m(\sigma) \qquad \sigma \in \Gamma_D \tag{15}$$

$$\sigma_{\xi\xi}(\sigma) + i\sigma_{\xi\eta}(\sigma) = 0 \qquad \sigma \in \Gamma_S \tag{16}$$

where u and v are the Cartesian components of the displacement, $\sigma_{\xi\xi}$ and $\sigma_{\xi\eta}$ denote stress components refered to the system (ξ, η) and ε represents the rotation of the inclusion.

Using the complex-variable formulation of the plane elastic problem for curvilinear boundaries[14] eqns (15) and (16) give the following equations

$$W_0^L(\sigma) - W_0^R(\sigma) = 0 \qquad \sigma \in \Gamma_S \tag{17}$$

$$\kappa W_0^L(\sigma) + W_0^R(\sigma) = 4i\mu\varepsilon m'(\sigma) \qquad \sigma \in \Gamma_D \tag{18}$$

for the complex function $W_0(\zeta) = m'(\zeta)W(\zeta)$ where $W(\zeta)$ is the usual complex potential of the theory of elasticity. $W_0^L(\sigma)$ and $W_0^R(\sigma)$ denote the limit values of $W_0(\zeta)$ as ζ tends to σ from L and R respectively, where L represents the image region of the matrix and R the circular hole in the ζ-plane. Equations (17) and (18) constitute a nonhomogeneous Hilbert problem with line of discontinuity the arc Γ_D described in a clockwise sense. It is obtained for the unknown function $W_0(\zeta)$ (Ref. 14)

$$W_0(\zeta) = \frac{4i\mu\varepsilon}{\kappa} X(\zeta)I(\zeta) + X(\zeta)R(\zeta) \tag{19}$$

where

$$I(\zeta) = \frac{1}{2\pi i} \int_{\Gamma_D} \frac{m'(\zeta)d\sigma}{X^L(\sigma)(\sigma - \zeta)} \tag{20}$$

$$R(\zeta) = (A_1\zeta + A_o) + \frac{A_{-1}}{\zeta} + \frac{A_{-2}}{\zeta^2} + \cdots + \frac{A_{-(\rho+1)}}{\zeta^{\rho+1}} \tag{21}$$

$$X(\zeta) = \frac{(\zeta - \sigma_2)^{\tau-1}}{(\zeta - \sigma_1)^{\tau}} \tag{22}$$

$$\tau = 0.5 + i\lambda, \quad \lambda = \frac{\log \kappa}{2\pi} \tag{23}$$

The function $X(\zeta)$ is the Plemelj function of the problem and is holomorphic in the whole ζ-plane cut along the arc Γ_D on which $X^R(\sigma) = -\kappa X^L(\sigma)$. Only the branch of the function $X(\zeta)$ for which $\lim_{\zeta \to \infty}[\zeta X(\zeta)] = 1$ will be considered in the sequel.

In calculating the integral $I(\zeta)$ it is observed that the function $F(\zeta) = m'(\zeta)/X(\zeta)$ is holomorphic in the whole plane cut along Γ_D except at the points $\zeta = \infty$ and $\zeta = 0$ at which it has poles of orders 1 and $(\rho + 1)$, respectively. Thus, the principal parts of $F(\zeta)$, $g_1(\zeta)$ and $g_2(\zeta)$ at $\zeta = \infty$ and $\zeta = 0$ will have the form

$$g_1(\zeta) = S_1\zeta + S_0 \tag{24}$$

$$g_2(\zeta) = \frac{S_{-1}}{\zeta} + \frac{S_{-2}}{\zeta} + \cdots + \frac{S_{-(\rho+1)}}{\zeta^{\rho+1}} \tag{25}$$

Using a contour Λ surrounding the arc Γ_D (Ref. 14) and observing that on $\Gamma_D X^R(\sigma) = -\kappa X^L(\sigma)$ the following is obtained for the integral $I(\zeta)$

$$I(\zeta) = \frac{\kappa}{2\pi i(1+\kappa)} \int_\Lambda \frac{F(\zeta)d\sigma}{\sigma - \zeta} \tag{26}$$

which by the well-known properties of the Cauchy integral gives

$$I(\zeta) = \frac{\kappa}{1+\kappa}\left[\frac{m'(\zeta)}{X(\zeta)} - g_1(\zeta) - g_2(\zeta)\right] \tag{27}$$

Introducing this value of $I(\zeta)$ into eqn. (19) we obtain for the function $W_0(\zeta)$

$$W_0(\zeta) = \mathscr{E}[m'(\zeta) - [g_1(\zeta) + g_2(\zeta)]X(\zeta)] + R(\zeta)X(\zeta) \tag{28}$$

where

$$\mathscr{E} = \frac{4i\mu\varepsilon}{1+\kappa} \tag{29}$$

Equation (29) gives a closed form solution for the unknown function $W_0(\zeta)$ from which the stress and displacement field are determined.

3.2. Determination of the Unknown Coefficients

The unknown coefficients $A_1, A_0, A_{-1}, A_{-2}, \ldots, A_{-(\rho+1)}$ of the function $R(\zeta)$ and the rotation ε of the inclusion are determined from the conditions that the complex potentials of the problem should be holomorphic in the region L of the ζ-plane and have a particular behavior at infinity. After lengthy calculations there is obtained[15]

$$A_0 = -RD_2\left(\frac{N_\infty + T_\infty}{2} + \frac{4i\mu\varepsilon_\infty}{1+\kappa}\right), \quad A_1 = -\frac{A_0}{D_2} \tag{30}$$

where ε_∞ is the rotation at infinity, while the $(\rho + 1)$ coefficients $A_{-1}, A_{-2},$ $\ldots, A_{-(\rho+1)}$ are determined from the following $(\rho + 1)$ equations

$$\bar{B}_{\rho+1} - \rho K_\rho = 0$$
$$\bar{B}_\rho - (\rho - 1)K_{\rho-1} = 0 \tag{31}$$

$$\cdots\cdots\cdots \quad \cdots\cdots\cdots$$

$$\bar{B}_3 - 2K_2 = 0$$
$$\bar{B}_2 - K_1 = R(N_\infty - T_\infty)e^{-2i\phi}$$
$$\bar{B}_1 = 0$$

$$\bar{B}_0 + K_{-1} = -\frac{iM + N'}{\pi R} - Rb_1(N_\infty - T_\infty)e^{-2i\phi}$$

where M is the moment of the stresses applied on A about the origin, N' is a real constant, and

$$\bar{B}_m = \sum_{s=0}^{\rho-m+1} A_{-(m+s)}d_s \tag{32a}$$

$$B_0 = \mathscr{E}\left(R - \sum_{s=0}^{\rho+1} S_{-s}d_s\right) + \sum_{s=0}^{\rho+1} A_{-s}d_s \tag{32b}$$

$$K_m = \sum_{s=0}^{\rho-m} T_{\rho-s}\alpha_{\rho-m-s}(m = -1, 0, 1, \ldots, \rho) \tag{33}$$

$$T_\rho = \bar{b}_\rho$$
$$T_{\rho-1} = \bar{b}_{\rho-1} \tag{34}$$

$$T_{\rho-m} = \bar{b}_{\rho-m} + \sum_{k=1}^{m-1} kb_k T_{\rho-m+1+k}(m = 2, 3, \ldots, (\rho+1))$$

with

$$b_{-1} = 1 \qquad b_0 = 0$$

$$\alpha_m = -\mathscr{E}\left[(m-1)Rb_{m-1} + \sum_{s=1}^{m+1} S_{s-m}D_s\right] + \sum_{s=1}^{m+1} A_{s-m}D_s \tag{35}$$

$$D(\omega) = \frac{\omega}{(1 - \omega\sigma_1)^{\tau}(1 - \omega\sigma_2)^{\bar{\tau}}} \tag{36a}$$

$$D_{\rho+1} = \frac{D^{(\rho+1)}(0)}{(\rho+1)!} = \frac{(\rho+1)D_1^{(\rho)}(0)}{(\rho+1)!} = \frac{D_1^{(\rho)}(0)}{\rho!} \tag{36b}$$

$$D_1^{(\rho+1)}(0) = \sum_{k=0}^{\rho} \binom{\rho}{k} D_1^{(\rho-k)}(0)\lambda^{(k)}(0)$$

$$\lambda(\omega) = \frac{\tau\sigma_1}{1 - \omega\sigma_1} + \frac{\bar{\tau}\sigma_2}{1 - \omega\sigma_2} \tag{37}$$

$$d_k = \frac{X^{(k)}(0)}{k!} \tag{38}$$

$$X^{(\rho+1)}(0) = \sum_{k=0}^{\rho} \binom{\rho}{k} X^{(\rho-k)}(0)\psi^{(k)}(0) \tag{39a}$$

$$X(0) = e^{\lambda(2\pi - \omega) - i\theta_0} \tag{39b}$$

$$\psi^{(k)}(0) = 2(-1)^{k+1} k! r \cos\left[\beta + \frac{k+1}{2}(\omega - 2\pi)\right] e^{-i(k+1)\theta_0} \tag{40}$$

$$r = \sqrt{0{\cdot}25 + \lambda^2}, \quad \beta = \tan^{-1} 2\lambda, \quad 0 \leqslant \beta < 2\pi \tag{41a}$$

$$\omega = \theta_2 - \theta_1, \quad \theta_0 = \frac{\theta_1 + \theta_2}{2} \tag{41b}$$

Furthermore, the coefficients of the functions $g_1(\zeta)$ and $g_2(\zeta)$ defined by eqns (24) and (25) are given by

$$S_1 = R, \quad S_0 = -RD_2 \tag{42}$$

and

$$\pi \cdot \mathbf{f} = \mathbf{c} \tag{43}$$

with

$$\mathbf{c} = [0 - Rb_1 - 2Rb_2 \ \ldots \ -\rho Rb_\rho]^T$$

$$\mathbf{f} = [S_{-1} S_{-2} \ \ldots \ S_{-(\rho+1)}]^T \tag{44}$$

$$\pi = \begin{bmatrix} d_0 & d_1 & \ldots & d_\rho \\ 0 & d_0 & \ldots & d_{\rho-1} \\ 0 & 0 & \ldots & d_0 \end{bmatrix}$$

3.3. Local Stress Distribution

The analysis of the stress distribution in the neighborhood of the crack tip is of particular importance and allows study of the growth character- istics of the interfacial crack. Consider the tip t_2 of the crack and the tangent $t_2 x$ of the inclusion at the point t_2 (Fig. 7). Using the asymptotic expansion

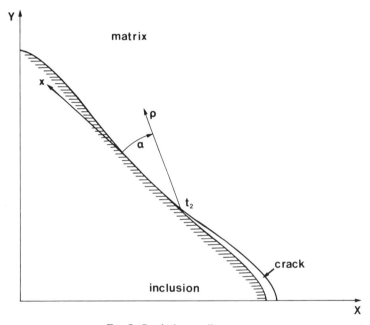

FIG. 7. Crack tip coordinate system.

of the function $W_0(\zeta)$ and its analytic continuation for $|\zeta| < 1$ around the point t_2 in conjunction with the equations relating $W_0(\zeta)$ to the stress field and after lengthy algebra the following equations of the curvilinear stress components $\sigma_{\xi\xi}$, $\sigma_{\eta\eta}$ and $\sigma_{\xi\eta}$ are obtained

$$\sigma_{\xi\xi} = \frac{e^{\lambda\alpha}}{2\sqrt{2\pi\rho}}\left[\cos\left(\frac{\alpha}{2} + \lambda\ln 2\pi\rho\right) + e^{2\lambda(\pi-\alpha)}\cos\left(\frac{\alpha}{2} - \lambda\ln 2\pi\rho\right)\right.$$

$$\left. + \sqrt{1+4\lambda^2}\sin\alpha\sin\left(\frac{3\alpha}{2} + \lambda\ln 2\pi\rho - \varphi\right)\right]K_1$$

$$+ \frac{e^{\lambda\alpha}}{2\sqrt{2\pi\rho}} \left[\sin\left(\frac{\alpha}{2} + \lambda\ln 2\pi\rho\right) - e^{2\lambda(\pi-\alpha)} \sin\left(\frac{\alpha}{2} - \lambda\ln 2\pi\rho\right) \right.$$

$$\left. - \sqrt{1+4\lambda^2} \sin\alpha\cos\left(\frac{3\alpha}{2} + \lambda\ln 2\pi\rho - \varphi\right) \right] K_2 \tag{45}$$

$$\sigma_{\eta\eta} = \frac{e^{\lambda\alpha}}{2\sqrt{2\pi\rho}} \left[3\cos\left(\frac{\alpha}{2} + \lambda\ln 2\pi\rho\right) - e^{2\lambda(\pi-\alpha)} \cos\left(\frac{\alpha}{2} - \lambda\ln 2\pi\rho\right) \right.$$

$$\left. - \sqrt{1+4\lambda^2} \sin\alpha\sin\left(\frac{3\alpha}{2} + \lambda\ln 2\pi\rho - \varphi\right) \right] K_1$$

$$+ \frac{e^{\lambda\alpha}}{2\sqrt{2\pi\rho}} \left[3\sin\left(\frac{\alpha}{2} + \lambda\ln 2\pi\rho\right) + e^{2\lambda(\pi-\alpha)} \sin\left(\frac{\alpha}{2} - \lambda\ln 2\pi\rho\right) \right.$$

$$\left. + \sqrt{1+4\lambda^2} \sin\alpha\cos\left(\frac{3\alpha}{2} + \lambda\ln 2\pi\rho - \varphi\right) \right] K_2 \tag{46}$$

$$\sigma_{\xi\eta} = \frac{e^{\lambda\alpha}}{2\sqrt{2\pi\rho}} \left[\sin\left(\frac{\alpha}{2} + \lambda\ln 2\pi\rho\right) - e^{2\lambda(\pi-\alpha)} \sin\left(\frac{\alpha}{2} - \lambda\ln 2\pi\rho\right) \right.$$

$$\left. + \sqrt{1+4\lambda^2} \sin\alpha\cos\left(\frac{3\alpha}{2} + \lambda\ln 2\pi\rho - \varphi\right) \right] K_1$$

$$- \frac{e^{\lambda\alpha}}{2\sqrt{2\pi\rho}} \left[\cos\left(\frac{\alpha}{2} + \lambda\ln 2\pi\rho\right) + e^{2\lambda(\pi-\alpha)} \cos\left(\frac{\alpha}{2} - \lambda\ln 2\pi\rho\right) \right.$$

$$\left. - \sqrt{1+4\lambda^2} \sin\alpha\sin\left(\frac{3\alpha}{2} + \lambda\ln 2\pi\rho - \varphi\right) \right] K_2 \tag{47}$$

$$\varphi = \tan^{-1} 2\lambda \tag{48}$$

The stress intensity factors K_1 and K_2 are given by

$$K_1 = e^{-\lambda(\omega/2)} \sqrt{\frac{\pi}{\sin\frac{\omega}{2}}} (A\cos\theta + B\sin\theta) \tag{49}$$

$$K_2 = e^{-\lambda(\omega/2)} \sqrt{\frac{\pi}{\sin\frac{\omega}{2}}} (A\sin\theta - B\cos\theta)$$

where

$$\theta = \theta_0 + \frac{2\pi + \omega}{4} + \lambda \ln\left[4\pi \sin \frac{\omega}{2} |m(\sigma_2)| \right], \quad \omega = \theta_2 - \theta_1, \quad 2\theta_0 = \theta_1 + \theta_2$$

$$A = \frac{\mathrm{Re}[m'(\sigma_2)]\mathrm{Re}[P(\sigma_2)] + \mathrm{Im}[m'(\sigma_2)]\,\mathrm{Im}[P(\sigma_2)]}{|m'(\sigma_2)|^{3/2}} \tag{50}$$

$$B = \frac{\mathrm{Re}[m'(\sigma_2)]\,\mathrm{Im}[P(\sigma_2)] - \mathrm{Im}[m'(\sigma_2)]\,\mathrm{Re}[P(\sigma_2)]}{|m'(\sigma_2)|^{3/2}}$$

with

$$P(\zeta) = R(\zeta) - \frac{4\mu\varepsilon i}{1+\kappa}[g_1(\zeta) + g_2(\zeta)] \tag{51}$$

3.4. The Square Inclusion

A rigid curvilinear rounded-off angle square inclusion with a symmetrically located interfacial crack is embedded in a plate subjected to a uniform biaxial stress system N and T at infinity (Fig. 8). For a critical value of the applied loads unstable crack extension takes place. The angle of initial crack extension α_0 is determined by assuming that the crack grows in the direction of the maximum circumferential stress σ_θ, while the critical

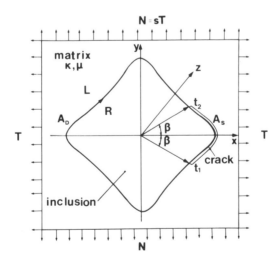

FIG. 8. Geometry of a rounded-off angle square inclusion partially bonded to an elastic matrix.

load is determined by

$$\sqrt{2\pi\rho}\,\sigma_\theta(\alpha_0) = K_m \qquad (52)$$

where K_m is the critical stress intensity factor which is a material constant.

The variation of the crack extension angle α_0 versus the crack angle β for $s = -1$, 0 and 1 is shown in Fig. 9, while Fig. 10 gives the dimensionless quantity $\bar{K}_m = K_m/(T_{cr}/\sqrt{\pi R})$. From Fig. 10 the critical stress T_{cr} for crack growth is determined.

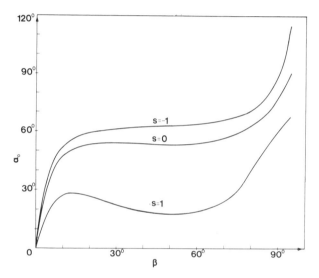

FIG. 9. Variation of the crack extension angle α_0 versus half crack angle β for a square inclusion for $s = -1$, 0 and 1.

3.5. The Triangular Inclusion

A rigid curvilinear rounded-off angle triangular inclusion with two locations of the interfacial crack shown in Fig. 11 is considered. Using the minimum strain energy density criterion it was found that for the case of Fig. 11(a) the crack always grows from its tip A. The variation of the dimensionless quantity $\bar{S}_{min} = 4 \cdot 17 \mu S_{cr}/(T_{cr}^2 R)$ versus half crack angle $\omega/2$ is shown in Fig. 12. Analogous results for the case of Fig. 11(b) where the crack again first starts from its tip A are shown in Fig. 13. From these figures the critical stress for unstable crack growth is determined.

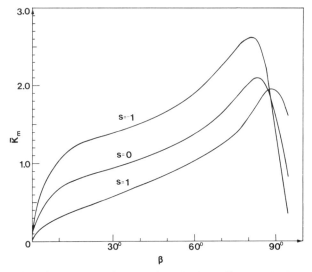

FIG. 10. Variation of the dimensionless stress intensity factor $\bar{K}_m$ versus half crack angle for a square inclusion for $s = -1$, 0 and 1.

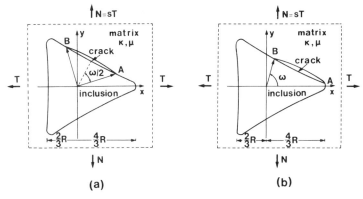

FIG. 11. A rounded-off angle triangular inclusion partially bonded to an elastic matrix. Geometrical configuration of two interfacial crack locations.

4. CONCLUDING REMARKS

The failure behavior of certain particulate composites consisting of filler particles with elastic moduli much higher than the elastic modulus of the matrix was studied. Two types of problems modelling the composite were

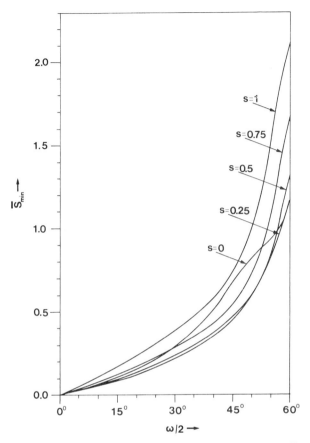

FIG. 12. Variation of the dimensionless minimum strain energy density factor $\bar{S}_{min}$ versus half crack angle $\omega/2$ for the case of Fig. 11(a). $s = 0$, 0·25, 0·50, 0·75 and 1·0.

considered: (i) rigid inclusions with cuspidal points embedded in an elastic matrix and (ii) rigid inclusions partially bonded to an elastic matrix. In the first case high stress concentrations are developed in the vicinity of the cuspidal points which constitute nuclei for failure initiation, while in the second case failure starts from the tips of the interfacial crack which coincides with the unbonded part of the inclusion. For the partially bonded inclusion a general solution of the stress and displacement fields was obtained for any inclusion shape using the method of complex potentials. The problem was reduced to a Hilbert problem and formulae for determining the unknown coefficients of the solution were derived. In both

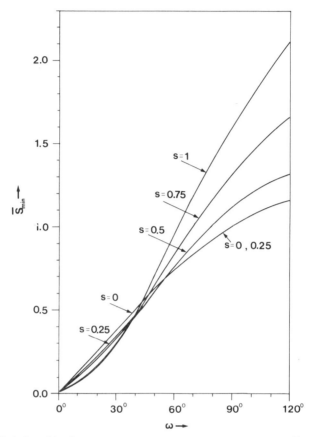

FIG. 13. Variation of the dimensionless minimum strain energy density factor $\bar{S}_{min}$ versus half crack angle $\omega/2$ for the case of Fig. 11(b). $s = 0$, 0.25, 0.50, 0.75 and 1.0.

cases results were obtained for special inclusion shapes including the fiber, the hypocycloidal, the astroidal, the square and the triangular inclusion. After determining the stress field a failure analysis of the composite took place using the maximum circumferential stress criterion and the strain energy density theory. The critical load for fracture initiation from the more vulnerable failure sites and the initial fracture angle were determined.

The results of this work shed light into the complicated problem of modelling the microstructure of particulate composites whose reinforcing constituents are of irregular shape like the various inorganic fillers, the metal or boron filaments, the aggregate or sand particles in concrete. In such case failure of the composite usually initiates from the sharp angles of the inclusions or the debonding areas of the different phases. The analysis of these types of failure mechanisms is of major importance for the understanding of the failure mode of the composite.

REFERENCES

1. PANASYUK, V. V., BEREZHNITSKII, L. T. and TRUSH, I. I., Stress distribution about defects such as rigid sharp-angled inclusions, *Problemy Prochnosti,* **7** (1972), 3–9.
2. ENGLAND, A. H., An arc crack around a circular inclusion, *J. Appl. Mech.,* **33** (1966), 637–40.
3. PERLMAN, A. B. and SIH, G. C., Elastostatic problems of curvilinear cracks in bonded dissimilar materials, *Int. J. Engng Sci.* **5** (1967), 845–67.
4. TOYA, M., A crack along the interface of a circular inclusion embedded in an infinite solid, *J. Mech. Phys. Solids,* **22** (1974), 325–48.
5. SUNDSTRÖM, B., An energy condition for initiation of interfacial microcracks at inclusions, *Engng Fract. Mech.,* **6** (1974), 483–92.
6. TOYA, M., Debonding along the interface of an elliptic rigid inclusion, *Int. J. Fracture,* **11** (1975), 989–1002.
7. VIOLA, E. and PIVA, A., Two arc cracks around a circular rigid inclusion, *Meccanica,* **15** (1980), 166–76.
8. VIOLA, E. and PIVA, A., Fracture behaviour by two cracks around an elliptic rigid inclusion, *Engng Fract. Mech.,* **15** (1981), 303–25.
9. SENDECKYJ, G. P., Debonding of rigid curvilinear inclusions in longitudinal shear deformation, *Engng Fract. Mech.,* **6,** (1974), 33–45.
10. SENDECKYJ, G. P., Elastic inclusion problems in plane elastostatics, *Int. J. Sol. Struct.,* **6** (1970), 1535–43.
11. SIH, G. C., Strain energy density factor applied to mixed mode crack problems, *Int. J. Fracture,* **10** (1979), 305–21.
12. SIH, G. C., A special theory of crack propagation, in: *Mechanics of Fracture 1: Methods of Analysis and Solutions of Crack Problems,* (G. C. Sih, ed.), Leyden, Noordhoff, 1973, pp. XXI–XLV.
13. GDOUTOS, E. E., *Problems of Mixed Mode Crack Propagation,* The Hague, Martinus Nijhoff, 1984.
14. MUSKHELISHVILI, N. I., *Some Basic Problems of the Mathematical Theory of Elasticity,* translated by J. R. M, Radok, 4th edn, Leyden, Noordhoff, 1975.
15. GDOUTOS, E. E. and KATTIS, M. A., A rigid curvilinear inclusion partially bonded to an elastic matrix (in press).

3

Thermomechanical Behaviour of Composites

R. JONES

Structures Division, Aero Research Laboratories,
Melbourne, Australia

and

T. E. TAY and J. F. WILLIAMS

Department of Mechanical Engineering, Melbourne University,
Victoria, Australia

ABSTRACT

This paper derives the governing equations for the thermomechanical behaviour of composites. When the basic equations for the thermoelastic behaviour of solids were first derived in the nineteenth century several approximations were made. The effect of these assumptions are discussed and illustrated by the results of a simple laboratory test. The implications of this work on the analysis of impact damaged laminates are then discussed.

1. INTRODUCTION

The theory describing the coupling between mechanical deformation and thermal energy of an elastic body was first published in 1858 by Lord Kelvin.[1] In composite materials absorption of moisture also results in internal stresses and/or strains. The thermal environment may also interact with moisture. Indeed it is generally accepted that the diffusion of moisture and temperature are also coupled.[2,3] However, tests have not yet been standardized for the required experimental measurements. The current theories for composites reduce to the theory of classical thermoelasticity when moisture effects are ignored and the material is elastic. However, when deriving the expression for the rate of change of entropy, the theory of classical thermoelasticity assumes that the stiffness tensor is independent of temperature. For metals this assumption has been shown to lead to erroneous results.[4]

This paper presents a consistent formulation for the thermomechanical behaviour of composites and then outlines the relevant criteria for failure due to delamination damage.

2. BASIC EQUATIONS

The equations governing the thermomechanical behaviour of a solid body due to heating and external forces during a reversible process are given below.

2.1. The Constitutive Equation

$$\sigma_{ij} = C_{ijkl}\varepsilon_{kl} - \beta_{ij}(T - T_o) - \phi_{ij}(M - M_o) \tag{1}$$

Here C_{ijkl} is the stiffness tensor, β_{ij} and ϕ_{ij} are coefficients related to the thermal and moisture expansion coefficients of the body respectively whilst T_o and M_o are the reference temperature and moisture content respectively.

2.2. Conservation of Mass (continuity equation)

$$\dot{M} = \frac{\partial \zeta}{\partial t} + \frac{\partial \zeta V_i}{\partial x_i}$$
$$= \frac{D\zeta}{Dt} + \zeta V_{i,i} \tag{2}$$

Here $V_i = u_{i,t}$, M is the mass flux of moisture per unit volume and ζ is the 'instantaneous' density.

2.3. Conservation of Momentum (no body forces)

$$\zeta \frac{DV_i}{Dt} + \dot{M}V_i = \sigma_{ij,j} \tag{3}$$

2.4. Conservation of Energy

$$\dot{E} = \sigma_{ij}V_{i,j} - T\dot{S} \tag{4}$$

Here S is the entropy and E is the internal energy.

2.5. Rate of Change of Entropy

$$T\dot{S} = -h_{i,i} + \zeta q \dot{M} \tag{5}$$

Here q is the heat generated when 1 g of moisture is absorbed by 1 g of the material and h is the heat flux tensor.

With this formulation the specific heat C_v at $\dot{\varepsilon}_{ij} = \dot{M} = 0$, is defined as

$$\zeta C_v \dot{T} = -h_{i,i}/_{\varepsilon_{ij},M} \tag{6}$$

The remaining equations may be expressed in a convenient form by introducing the free energy, F, such that

$$F = E - TS \tag{7}$$

The quantity F can be expressed in terms of strain, temperature and moisture in a manner analogous to that given in Ref. 5, viz.

$$F = 1/2\varepsilon_{ij} C_{ijkl} \varepsilon_{kl} - \beta_{ij} \varepsilon_{ij} (T - T_o) - \phi_{ij} \varepsilon_{ij} (M - M_o) + C_1(T, M) \tag{8}$$

Here C_1 is a function of temperature and moisture and as shown in Ref. 5

$$\zeta C_v = -T \frac{\partial^2 F}{\partial T^2}$$

$$= -T \frac{\partial^2 C_1}{\partial T^2}\Big/_{\varepsilon_{ij},M} \tag{9}$$

The Duhamel–Neuman law tells us that

$$\sigma_{ij} = \left(\frac{\partial F}{\partial \varepsilon_{ij}/_{T,M}} \right) \tag{10}$$

whilst

$$S = -\frac{\partial F}{\partial T/_{\varepsilon_{ij},M}}$$

$$= -\varepsilon_{ij} \frac{\partial C_{ijkl}}{\partial T} \varepsilon_{kl} + \varepsilon_{ij} \frac{\partial}{\partial T} (\beta_{ij}(T - T_o))$$

$$+ \varepsilon_{ij} \frac{\partial}{\partial T} (\phi_{ij}(M - M_o)) - \frac{\partial C_1}{\partial T} \tag{11}$$

With this notation the change in entropy can be written as

$$\dot{S} = \frac{\partial S}{\partial T} \dot{T} + \frac{\partial S}{\partial M} \dot{M} + \frac{\partial S}{\partial \varepsilon_{ij}} \dot{\varepsilon}_{ij} \tag{12}$$

Now
$$\frac{\partial S}{\partial T} = -\frac{\partial^2 C_1}{\partial T^2} = \zeta C_v/T \qquad (13)$$

and
$$\frac{\partial S}{\partial \varepsilon_{ij}} = -\frac{\partial C_{ijkl}}{\partial T}\varepsilon_{kl} + \frac{\partial}{\partial T}(\beta_{ij}(T-T_o)) + \frac{\partial}{\partial T}(\phi_{ij}(M-M_o)) \qquad (14)$$

whilst
$$\frac{\partial S}{\partial M} = -\frac{\partial^2 F}{\partial M \partial T} = 0 \qquad (15)$$

Substituting eqns (13), (14) and (15) into (12) we finally obtain the 'Entropy Equation'

$$T\dot{S} = \zeta C_v \dot{T} + \dot{\varepsilon}_{ij} T\left(\frac{\partial}{\partial T}(\beta_{ij}(\dot{T}-T_o)) + \frac{\partial}{\partial T}(\phi_{ij}(M-M_o))\right.$$
$$\left. -\frac{\partial}{\partial T}(C_{ijkl})\varepsilon_{kl}\right) \qquad (16)$$

If we now substitute eqn. (5) into (16) we obtain

$$-h_{i,i} + \zeta q\dot{M} = \zeta C_v \dot{T}$$
$$-\left(\varepsilon_{kl}\frac{\partial C_{ijkl}}{\partial T} - \frac{\partial}{\partial T}(\beta(T-T_o))\right.$$
$$\left. -\frac{\partial}{\partial T}(\phi_{ij}(M-M_o))\right)T\dot{\varepsilon}_{ij} \qquad (17)$$

Fourier's law for heat conduction may be used in most circumstances and the term $h_{i,i}$ occurring in eqn. (17) can be replaced by

$$-\frac{\partial}{\partial x_i}\left(k_{ij}\frac{\partial T}{\partial x_j}\right)$$

where k_{ij} is the thermal conductivity tensor.

In order to solve the above set of equations it is necessary to specify an equation describing the way in which moisture is absorbed into a composite. Most formulations are empirical. The equation preferred by the authors was first derived in Ref. 6 and is given below, viz.

$$\frac{\partial}{\partial x_i}\left(D_{ij}\frac{\partial M}{\partial x_j}\right) = \frac{\partial}{\partial t}(M-\lambda T) \qquad (18)$$

where D_{ij} and λ are material constants. An extension to allow for mechanical coupling in the matrix material replaced the term T on the right hand side of eqn. (18) by a term of the form $T + N\psi$ where N is an experimental constant and

$$\psi = \sigma_{kk} + \beta(T - T_o) + \phi(M - M_o) \tag{19}$$

Here β and ϕ are thermal and moisture expansion coefficients for the matrix material.

3. IMPLICATIONS FOR CYCLIC STRESSING

If an elastic body is subjected to cyclic stresses and the frequency is such that the process is adiabatic then, from eqns (5) and (16), it follows that

$$\frac{\dot{T}}{T} = \left(\varepsilon_{kl} \frac{\partial C_{klij}}{\partial T} - \frac{\partial}{\partial T} (\beta_{ij}(T - T_o)) - \frac{\partial}{\partial T} (\phi_{ij}(M - M_o)) \right) \dot{\varepsilon}_{ij} / \zeta C_v \tag{20}$$

Most papers neglect the term $\partial C_{klij}/\partial T$.

If this is done $\dot{\varepsilon}_{ij}$ is said to be related to $\dot{T}$ by a material constant. However, it is well documented[7,8] that even for metals this constant is stress/strain dependent.

As shown in Ref. 4 the present theory is able to accurately predict the stress dependency of the thermoelastic constant K for both a titanium and an aluminium alloy. The thermoelastic constant and the Gruneisen parameter for a metal are related by the formulae

$$K = (1 - 2v) \, \gamma / E \tag{21}$$

In order to illustrate this effect let us consider a metal bar subjected to a uniaxial stress

$$\sigma_{11} = S_m + \Delta S \sin \omega t, \quad \sigma_{ij} = 0 \text{ if } i, j \neq 1 \tag{22}$$

where S_m is the mean stress.

Substituting eqn. (22) into (20) gives

$$\zeta C_v \frac{\dot{T}}{T} = - \left(\alpha - \frac{1}{E^2} \frac{\partial E}{\partial T} S_m \right) \omega \Delta S \cos \omega t + \frac{1}{2E^2} \frac{\partial E}{\partial T} \omega (\Delta S)^2 \sin 2\omega t \tag{23}$$

Integrating with respect to t we obtain

$$\frac{\Delta T}{T_o} = - \left(\alpha - \frac{1}{E^2} \frac{\partial E}{\partial T} S_m \right) \frac{\Delta S}{\zeta C_v} \sin \omega t$$

$$+ \frac{1}{4E^2 \zeta C_v} \frac{\partial E}{\partial T} (\Delta S)^2 (1 - \cos 2\omega t) \tag{24}$$

where T_o is the reference temperature. From eqn. (24) we see that the thermoelastic constant, K, is given by

$$K = \left(\alpha - \frac{1}{E^2} \frac{\partial E}{\partial T} S_m \right) \Big/ \zeta C_v \tag{25}$$

and has a mean stress dependence. Furthermore if, as in Ref. 4, we define $K_o = \alpha/\zeta C_v$ then

$$\frac{1}{K_o} \frac{\partial K}{\partial S_m} = - \frac{1}{E^2} \frac{\partial E}{\partial T} \tag{26}$$

The quantity $(1/K_o)(\partial k/\partial S_m)$ was measured experimentally by Machin et al.[8] for a titanium alloy Ti-6Al-4V and an aluminium alloy Al-2024 for which the values $\partial E/\partial T$ were available.[4,9] Table 1 lists the data used and the comparison between the theoretically predicted mean stress dependence and the experimental results. Good agreement between theory and experiment is clearly evident.

A more detailed analysis and discussion of this problem is contained in Ref. 4.

TABLE 1
Comparison of theoretical and measured mean stress dependence of K (from Ref. 4)

Material	α ($°C^{-1}$)	E (MPa)	dE/dT (MPa/°C)	$(dK/dS_m)K_o^{-1}$ (MPa^{-1}) Theory	(MPa^{-1}) Experiments[8]
Ti-6Al-4V	9.0×10^{-6}	1.11×10^5	-48.0	4.33×10^{-4}	4.29×10^{-4}
Al-2024	2.3×10^{-5}	7.2×10^4	-36.0	3.02×10^{-4}	3.19×10^{-4}

For composite materials the change of the stiffness tensor with temperature[10] is far greater than for metals, see Table 2, and should not be ignored.

TABLE 2
Graphite-epoxy (AS/3501-6) orthotropic elastic properties (from Ref. 10)

Temperature (°F)	Parameter ($lb/in^2 \times 10^{-6}$)				
	ν_{12}	E_1^T	E_1^C	E_2^T and E_2^C	G_{12}
−65 (dry)	0.3	18.70	17.60	2.30	1.00
RT (dry)	0.3	18.70	17.60	1.90	0.85
220 (dry)	0.3	18.40	17.30	1.65	0.65
250 (dry)	0.3	18.40	17.20	1.60	0.63
220 (wet)	0.3		17.30	1.32	0.52

The temperature rise due to mechanical loading is usually small (i.e. of the order $1/K$) unless there is extensive plasticity. However in the vicinity of a crack or a delamination ε_{ij} is infinite and eqn (16) predicts that the glass transition temperature will be exceeded in a small localized region. However the validity of this approach which is based on reversible thermodynamics is questionable once crack extension or delamination growth has occurred.

The residual thermal strains ε_{ij}^{T} due to the curing process may also play a role in the variability of fatigue life. The residual strains may be interpreted as the mean strain level about which the mechanical strains oscillate. As shown above the temperature depends both on the mean stress/strain and the cyclic stress/strain fields.

4. FRACTURE MECHANICS

There are two fracture parameters which are widely used to predict the residual strength of delaminated composite laminates. These are:

(i)   the energy release rate approach;
(ii) the strain energy density approach.

4.1. The Energy Release Rate Approach

For mode 1 self similar crack growth of a through crack in the absence of body forces the energy release rate G can be written as

$$G = \lim_{\varepsilon \to 0} \int_{\Gamma_\varepsilon} \left(W \cdot n_1 - t_i \frac{\partial u_i}{\partial x_1} \right) ds \tag{27}$$

where W is the energy density, Γ_ε is a vanishing small closed path around the tip with normal $\mathbf{n}$ and t_i are the components of the traction vector on the path.

Here W is defined as

$$W = 1/2 \varepsilon_{kl} C_{ijkl} \varepsilon_{ij} - \beta_{ij} \varepsilon_{ij} (T - T_o) - \phi_{ij} \varepsilon_{ij} (M - M_o) + C_1(T, M) \tag{28}$$

Let us now consider the integral J_s which we will define as

$$J_s = \int_{\Gamma_s} \left(W n_1 - t_i \frac{\partial u_i}{\partial x_1} \right) ds \tag{29}$$

where Γ_s is the external boundary of the body. As mentioned in Ref.11 there is a tendency to drop the subscript s and refer to J_s as J. Using

Green's theorem it follows that

$$G = J_s - \int_{V_s - V_\varepsilon} \left(\frac{\partial W}{\partial x_1} - \frac{\partial}{\partial x_j} \left(\sigma_{ij} \frac{\partial u_i}{\partial x_1} \right) \right) dv \qquad (30)$$

where $V_s - V_\varepsilon$ is the area between the curves Γ_ε and Γ_s. If we assume that β_{ij}, ϕ_{ij} and C_{ijkl} are only functions of T and M then the term $\partial w/\partial x_1$ can be written as

$$\frac{\partial W}{\partial x_1} = \frac{\partial}{\partial x_j} \left(\sigma_{ij} \frac{\partial u_i}{\partial x_1} \right) + \frac{1}{2} \varepsilon_{ij} \frac{\partial C_{ijkl}}{\partial x_1} \varepsilon_{kl}$$

$$- \varepsilon_{ij} \frac{\partial}{\partial x_1} [\beta_{ij}(T - T_0) + \phi_{ij}(M - M_0)] + \frac{\partial C_1}{\partial x_1} \qquad (31)$$

and

$$G = J_s - \int_{V_s - V_\varepsilon} \left[\frac{1}{2} \varepsilon_{ij} \frac{\partial C_{ijkl}}{\partial x_1} \varepsilon_{kl} - \varepsilon_{ij} \frac{\partial}{\partial x_1} (\beta_{ij}(T - T_0) \right.$$

$$\left. + \phi_{ij}(M - M_0)) + \frac{\partial C_1}{\partial x_1} \right] dV \qquad (32)$$

Thus J_s will not equal the energy release rate G unless the area integral vanishes. At constant moisture and temperature J_s may be equal to G. However in general the area integral will be non-zero and J_s, which is measured experimentally from the movement of the load points,[12] will not equal G.

In service aircraft heating is often localized and the moisture content varies. This will produce a spatial variation in the tensor C_{ijkl} with the result that the area integral will in general be non-zero. Consequently when designing laboratory tests care should be taken to reproduce the near tip stress, strain, temperature and moisture fields rather than reproducing the 'global behaviour'. This is particularly true if the aim of the test is to establish such quantities as the critical damage size or the maximum permissible load.

For a three-dimensional fracture problem the integral on the right hand side of eqn. (30) is no longer equal to the energy release rate and is referred to as T^*.

5. STRAIN ENERGY DENSITY

In the strain energy density approach failure is assumed to occur when the available energy density W_{av} at a distance r_o in front of the delamination in the direction of growth reaches a critical value W_c. The value W_c is dependent both on the values of $dV(=\varepsilon_{11} + \varepsilon_{22} + \varepsilon_{33})$ and dA, the area of crack growth.

For thermomechanical problems

$$W_{av} = 1/2\sigma_{ij}\varepsilon_{ij} - 1/2\sigma_{ij}\alpha_{ij}(T - T_o) - 1/2\sigma_{ij}\psi_{ij}(M - M_o) - W_f \qquad (33)$$

where $\beta_{ij} = \alpha_{ij}C_{ijkl}$, $\phi_{ij} = \psi_{ij}C_{ijkl}$ and W_f is the energy density in the fibre.

As an illustration of this approach consider an impact damaged laminate with a fastener hole under compression. The dimensions of the model are the same as those used in the experimental work of Ref. 13, see Fig. 1 in this reference. The specimen tested was a $[0/45/0_2/-45/0_2/-45/0]_s$ T300/5208 graphite-epoxy laminate and contained a centrally located hole 9·5 mm in diameter, surrounded by delamination damage due to impact and poor drilling. The elements used are mostly twenty-noded isoparametric elements with directionally reduced integration and $2 \times 2 \times 3$ Gaussian quadrature points, with the 3 points being taken through the ply thickness. The crack tip elements along the circular delamination are fifteen-noded isoparametric wedge elements.

The initial damage around the fastener hole (from Ref. 13) is modelled as a circular delamination of radius 13·75 mm between the second and third plies (i.e. between 45° and 0° plies). It can be seen from the ultrasonic C-scan that the initial delamination is nearly circular.[13]

The two plies above and below the delamination and the matrix region around the delamination are modelled separately with ordinary three-dimensional elements while the remaining 20 plies are modelled with super-elements with displacements varying quadratically in the local isoparametric co-ordinate system. The material properties used are those of AS1/3501–6.

It is important that in the FE model, the faces of the delamination are prevented from overlapping. Otherwise, non-physical solutions may be obtained. By examining the solutions of the displacements, it is found that some parts of the delaminated faces have overlapped. Thus a series of constraint equations are applied to appropriate nodes to simulate local closure.

Two load cases are considered, viz.

1. A compressive load of 150 kN applied at the ends of the model.
2. The load case mentioned above together with a 1·0% moisture content. To induce stresses due to moisture the external boundary of the specimen is prevented from moving in the direction perpendicular to the primary load (i.e. load case 1). Table 3 shows the maximum value of T^* for the two load cases as well as the maximum values of the parameter W_t where

$$W_t = \lim_{\varepsilon \to 0} \int_{\Gamma_\varepsilon} W n_1 \, ds \qquad (34)$$

TABLE 3
*Maximum values of T^* and $W_t(J/M)$
around the delamination*

Load case	T^*	W_t
1	71·8	54·1
2	168·0	57·3

Here we see that the value of T^* is increased dramatically by the presence of moisture. However, for the present problem the interlaminar stresses σ_z, τ_{xz} and τ_{yz} are relatively unaffected by the moisture content. This phenomenon is also seen in the ratio of the maximum value of the strain energy density W in the matrix material directly in front of the delamination for the two load cases:

$$W(\text{load case } 1)/W(\text{load case } 2) = 1·02$$

Indeed from Table 3 we see that W_t, which involves the integral of the energy density around the delamination, is also relatively unaffected by the presence of moisture. This infers that, in the present problem, the presence of moisture does not significantly increase the likelihood of static failure by means of delamination growth along the interface. This observation is consistent with the experimental results given in Refs. 14 and 15. We also see that if the growth of the damage is likely to be non-self-similar then energy release rate methods must be used with extreme caution.

REFERENCES

1. THOMSON, W. On the thermoelastic and thermo-magnetic properties of matter, *Q. J. Maths*, **1** (1855), 55–77.

2. SPRINGER, G. S. Environmental effects on epoxy matrix composites, ASTM-STP 674, 1979, pp. 291–312.
3. SIH, G. C., Analysis of defects and damage in composites, *Proc. ICF6, Advances in Fracture Research*, (S. R. Valluri *et al.*, eds), Oxford, Pergamon Press 1984, pp. 525–48.
4. WONG, A. K., JONES, R. and SPARROW, J. G. Thermoelastic constant or thermoelastic parameter, submitted to *J. Physics and Chemistry of Solids*. (in press).
5. FUNG, Y. C., *Foundations of Solid Mechanics*, New Jersey, Prentice Hall, 1965.
6. SIH, G. C. and SHIH, M. T., A generalized theory of coupled hygrothermal elasticity: transient effects in composite laminate with circular cavity, AMMRC TR 83-56.
7. BELGEN, M. H., Infra red radiometric stress instrumentation application range study, NASA CR-1067, 1968.
8. MACHIN, A. S., SPARROW, J. G. and STIMSON, M. G., Mean stress dependence of the thermoelastic constant, *J. Strain*, (in press).
9. *Metals Handbook*, 9th edn, Vol. 3 American Society for Metals, 1981.
10. MURRAY, J. E. and BIRCHFIELD, E. B. Material substantiating data and analysis report, McDonnell Aircraft Company, MDC AS-253, 1978.
11. BRUST, F. W., McGOWAN, J. J. and ATLURI, S. N., A combined numerical/experimental study of ductile crack growth after a large unloading using T^*, J and CTOA criteria, *Engng Fract. Mech.*, **22** (3) (1986), 537–50.
12. ATLURI, S. N., NAKAGUKI, M., NISHIOKA, T. and KUANG, Z. B., Crack tip parameters and temperature rise in dynamic crack propagation, *Engng Fract. Mech.*, **23** (1), (1986), 167–82.
13. LAURAITIS, K. N., RYDER, T. and PETTIT, D. E., Advanced residual strength degradation rate modelling for advanced composite structures. Vol. II, AFWAL–TR–79–3095, 1981.
14. MOHLIN, T., CARLSON, L. F. and GUSTAVSSON, A. I., Delamination growth in a notched graphite epoxy laminate under compression fatigue loading, ASTM STP 876, 1985, pp. 168–188.
15. LAURATIS, K. N. and SANDORFF, P. E., Experimental investigation of the interaction of moisture, low temperature and low level impact on graphite/epoxy composites, NADC 79102–60, 1980.

4

Fracture Tests for Mixed Mode Failure of Composites Laminates

J. G. WILLIAMS

Mechanical Engineering Department, Imperial College of Science and Technology, London, UK

ABSTRACT

A method of calculating the energy release rate, G, for plane delaminations in composites is given. This is couched in terms of local moments and forces and is a useful general method of obtaining such solutions. A scheme for the exact partitioning of the G values into mode I and II components is also given. The method is then used to give results for both a constant ratio mixed mode test and one in which the ratio continuously varies. Data are presented for three matrix resins used in carbon fibre laminates and for two of them, epoxy and PEEK, a unique locus of G_I and G_{II} is defined. For bismaleimide fibre bridging results in G_c increasing with crack growth and consequent differences between the tests. The test methods are generally confirmed as satisfactory.

1. INTRODUCTION

Laminates of high strength and stiffness fibres with matrices of tough polymers are of considerable commercial interest since they offer a significant potential weight advantage over conventional materials. Their very nature results in a structure in which these properties are realised in the plane of the laminate but their weakness lies in the through thickness or translaminar direction. Here strength and stiffness are approximately those of the matrix. In designing structures care is taken to exploit the in-plane properties but inevitably some forms of loading can induce failure between laminate layers, i.e. delamination, and this is often the limiting property of the structure. It is therefore important to have a good understanding of delamination and to be able to characterise it under any loading system so that design predictions can be made. This paper describes a scheme by which such a characterisation may be carried out.

The methodology is essentially that of linear elastic fracture mechanics (LEFM) in that the structure behaves in a linear elastic fashion and failure is assumed to occur in a plane parallel to the surface. This is a very important simplification in the analysis since it avoids the necessity of predicting the *direction* of crack growth under complicated loading systems. It is assumed that the laminated structure is such that the crack propagation is forced to occur in the plane of the sheet. The loadings which effect this growth can be bending moments, shear forces or in-plane loads and may involve buckling of the sheet. In such cases the crack can be forced to grow under a combination of an opening mode (mode I), and a sliding mode (mode II). In isotropic homogeneous systems such loading can lead to non-colinear growth but here it is forced to be colinear. Under these circumstances we must seek a fracture criterion for mixed-mode crack propagation and to do this we must have tests which enable us to propagate cracks under mixed-mode conditions. In addition we must be able to analyse the loading condition so that we may determine the energy release rates for each mode. Such an analysis coupled with mixed-mode fracture criteria will provide a basis for a design method.

2. METHOD OF ANALYSIS

In this analysis we shall derive expressions for the energy release rate for the situation shown in Fig. 1 in which there is a single, through thickness, delamination in a laminate of thickness $2h$ which is located a distance h_1 from one surface and in which $h_1 + h_2 = 2h$, $h_1 < h_2$. This is essentially a one-dimensional model and enables the analysis to be conducted in terms

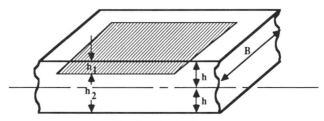

FIG. 1. Delamination.

of simple beam theory. Practical problems of other shapes could be tackled in a similar way but would lead to more complicated solutions. The general method is dealt with in detail elsewhere[1,2] and will be considered only in outline here. Let us consider one end of the delamination of length a as shown in Fig. 2. The general loadings at this point are a moment of

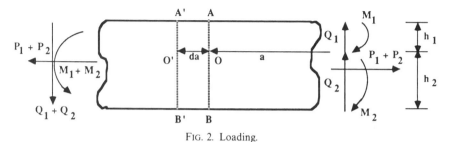

FIG. 2. Loading.

$M_1 + M_2$, a shear force of $Q_1 + Q_2$ and an axial load of $P_1 + P_2$. In the cracked section these loads are M_1, Q_1 and P_1 on the upper section of thickness h_1 and similarly M_2, Q_2, P_2 on h_2. Let us consider first only the moments and note that for a section of second moment of area I and axial modulus E the strain energy per unit length of the beam is given by $M^2/2EI$. In Fig. 3 we have the moment versus angle of rotation, ϕ, relationship for

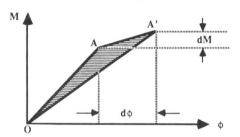

FIG. 3. Bending moment loading lines.

the section with a crack of length a, i.e. the line OA, which is linear in this case. If the crack grows to $(a + da)$ then the loading line becomes OA' and the change in energy of the system for this change in a is given by the shaded triangular area OAA'.

Now the strain energy $U_s = 1/2(M\phi)$ so G may be written as:

$$dU = \frac{1}{2}M\phi + \left(M + \frac{dM}{2}\right)d\phi - \frac{1}{2}(M + dM)(\phi + d\phi)$$

i.e. $$BG = \frac{dU}{da} = \frac{1}{2}\left(M\frac{d\phi}{da} - \phi\frac{dM}{da}\right)$$

$$\left.\begin{array}{r} BG = -\frac{1}{2}\phi\frac{dM}{da}\bigg|_{\phi\,\text{const}} = -\frac{dU_s}{da}\bigg|_{\phi\,\text{const}} \\[3mm] \text{or} \quad\quad BG = +\frac{1}{2}M\frac{d\phi}{da}\bigg|_{M\,\text{const}} = +\frac{dU_s}{da}\bigg|_{M\,\text{const}} \end{array}\right\}$$ (1)

The former is the more usual route but the latter is more convenient here. It should be noted that the use of constant ϕ or M in the analysis to find G does not imply that this is the loading in practice but is simply a device for calculation. Returning now to Fig. 2 we can compute dU_s/da at constant M here by calculating the energy change within the contour ABB′A when a increases by da (O to O′);

i.e. $$\frac{dU_s}{da} = \frac{1}{2}\frac{M_1^2}{EI_1} + \frac{1}{2}\frac{M_2^2}{EI_2} - \frac{1}{2}\frac{(M_1 + M_2)^2}{EI_0}$$

where $$I_1 = \frac{Bh_1^3}{12} = 8\xi^3 I, \quad I_2 = 8(1-\xi)^3 I, \quad I_0 = 8I$$

and $$\xi = \frac{h_1}{2h} \quad \text{with} \quad I = \frac{Bh^3}{12}$$

The expression for the total G now becomes,

$$G = \frac{1}{16BEI}\left[\frac{M_1^2}{\xi^3} + \frac{M_2^2}{(1-\xi)^3} - (M_1 + M_2)^2\right]$$ (2)

This is an important result since it enables G to be calculated from the *local* values of M, B, E and I of the crack tip without recourse to the values in remote regions. Similar analyses lead to expressions for G for P and Q;

$$G = \frac{1}{4BEA}\left[\frac{P_1^2}{\xi} + \frac{P_2^2}{(1-\xi)} - (P_1 + P_2)^2\right]$$ (3)

and $$G = \frac{3}{10B\mu A}\left[\frac{Q_1^2}{\xi} + \frac{Q_2^2}{(1-\xi)} - (Q_1 + Q_2)^2\right]$$ (4)

when $A = Bh$ and μ is the shear modulus.

These expressions are for the total energy release rate and it is necessary to partition them into modes I and II. In bending, mode II occurs when there is crack growth with the curvature in both sections equal. If we have a moment M_{II} on h_1 and ψM_{II} on h_2 then we have pure mode II when:

$$\frac{d\phi}{da} = \frac{M_{II}}{EI_1} = \frac{\psi M_{II}}{EI_2}$$

i.e.

$$\psi = \frac{I_2}{I_1} = \left(\frac{1-\xi}{\xi}\right)^3$$

Mode I requires equal moments in opposite senses so we may write;

$$M_1 = M_{II} - M_I \quad \text{and} \quad M_2 = \psi M_{II} + M_I$$

Substituting these expressions in eqn. (2) we have a term in M_{II}^2, one in M_I^2 but that in $M_I M_{II}$ is zero thus giving exact partitioning. We may thus write;

$$\left.\begin{array}{l} G_I = \dfrac{M_I^2}{BEI} \cdot \dfrac{(1+\psi)}{16(1-\xi)^3} = \dfrac{(M_2 - \psi M_1)^2}{BEI} \cdot \dfrac{1}{16(1-\xi)^3(1+\psi)} \\[3mm] G_{II} = \dfrac{M_{II}^2}{BEI} \cdot \dfrac{3}{16} \dfrac{(1-\xi)}{\xi^2}(1+\psi) = \dfrac{(M_2 + M_1)^2}{BEI} \cdot \dfrac{3}{16} \dfrac{(1-\xi)}{\xi^2(1+\psi)} \end{array}\right\} \quad (5)$$

Axial loads give only mode II so that; $G_I = 0$ and G_{II} is given by eqn. (3) and for shear forces $G_{II} = 0$ and G_I is given by eqn. (4).

3. MIXED MODE TESTS

The analysis of mixed mode crack propagation requires some form of criterion if it is to be employed in design. The most obvious would be that the total energy release rate remained the same, i.e. G constant, but this appears not to be so and there is usually a distinct difference between G_{IC} and G_{IIC}. The criterion can be represented as a locus of G_I versus G_{II} at fracture and some form of relationship,

$$F(G_I, G_{II}) = 0$$

is required. In order to explore this it is necessary to perform tests in various combinations of modes I and II and the following have been employed:

(1) pure mode I;
(2) pure mode II;

(3) mixed mode I/II in which the ratio G_I/G_{II} remained constant;

(4) mixed mode I/II in which G_I/G_{II} varied continuously.

These latter two enable the important characteristic of history dependence to be studied.

Figure 4 shows the test configurations used in this test series and the

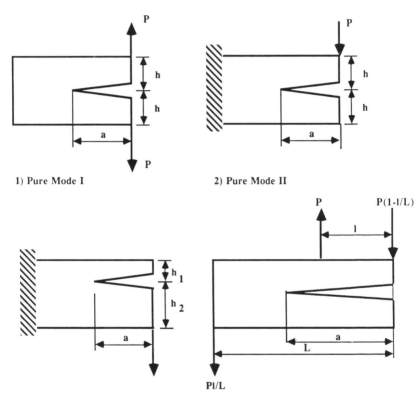

1) Pure Mode I   2) Pure Mode II

3) Fixed Ratio Mixed Mode   4) Variable Ratio Mixed Mode

FIG. 4. Test configurations.

analysis employed is as follows:

$$(1) \quad M_2 = M_1 = Pa, \quad \xi = \frac{1}{2}, \quad G_I = \frac{P^2 a^2}{BEI}, \quad G_{II} = 0 \tag{6}$$

$$(2) \quad M_2 = -M_1 = \frac{Pa}{2}, \quad \xi = \frac{1}{2}, \quad G_I = 0, \quad G_{II} = \frac{3}{16} \frac{P^2 a^2}{BEI} \tag{7}$$

(3) $M_1 = 0$, $M_2 = Pa$

$$G_I = \frac{P^2 a^2}{BEI} \frac{1}{16(1-\xi)^3(1+\psi)}, \quad G_{II} = \frac{P^2 a^2}{BEI} \frac{3(1-\xi)}{16\xi^2(1+\psi)} \quad (8)$$

$$\frac{G_I}{G_{II}} = \frac{3(1-\xi)^4}{\xi^2}, \text{ constant}$$

(4) $M_1 = \dfrac{Pl}{4}\left[1 + \left(\dfrac{1}{a}\right)^2 - 2\left(\dfrac{a}{L}\right)\right]$, $\quad M_2 = \dfrac{Pl}{4}\left[3 - \left(\dfrac{1}{a}\right)^2 - 2\left(\dfrac{a}{L}\right)\right]$

$$G_I = \frac{P^2 l^2}{16BEI} \cdot \left[1 - \left(\frac{1}{a}\right)^2\right]^2, \quad G_{II} = \frac{3P^2 l^2}{16BEI}\left[1 - \left(\frac{a}{L}\right)\right]^2 \quad (9)$$

$$\frac{G_I}{G_{II}} = \frac{1}{3}\left[\frac{1-(1/a)^2}{1-(a/L)}\right]^2$$

4. EXPERIMENTAL RESULTS

The experiments will be described in detail elsewhere [3,4] and only an outline of the results will be given here. All four test configurations were used on unidirectional laminates of carbon fibres using three different matrices, i.e. poly(ether-ether ketone) (PEEK), an epoxy and bismaleimide. In each test the cracks were grown by loading the specimens in an Instron testing machine and recording the load, load point displacement, and the crack length by observing marks on the specimen edge. G values can, of course, be found via eqns (6–9) from P and a only but the deflection, δ, is useful since for the mode I test with a uniform section EI can be found from the relationship;

$$\delta = \frac{2}{3} \frac{Pa^3}{EI}$$

and this value was used in all the subsequent tests.

During each test it was possible to evaluate G as the crack grew and Fig. 5 shows G as a function of crack growth for different modes of testing for PEEK. In mode I there is clearly a constant G value but in mode II there is an increase in G as a increases. There is an extreme example of this effect shown in Fig. 6 for bismaleimide and this gives some hint as to the source since adhesion of the fibres to the matrix here was not good (the fibres were not correctly primed). This results in the fracture occurring in several planes

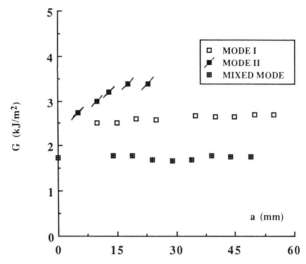

FIG. 5. Variation of *G* with crack growth for PEEK.

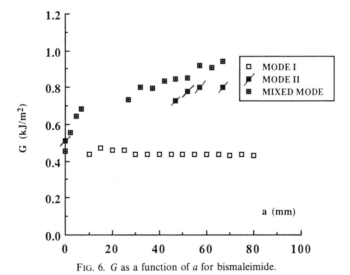

FIG. 6. *G* as a function of *a* for bismaleimide.

and in fibres 'bridging' the two arms giving higher values which increase as the crack grows. In the PEEK and epoxy specimens the effect is much less marked because the adhesion is better.

Results for the three materials as G_I versus G_{II} are shown in Figs 7, 8 and

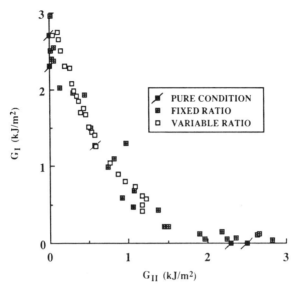

FIG. 7. Failure locus under mixed mode loading for PEEK.

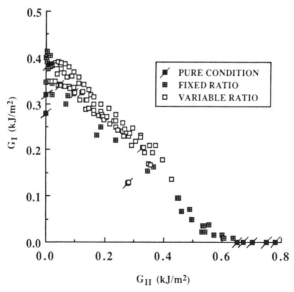

FIG. 8. Failure locus under mixed mode loading for epoxy.

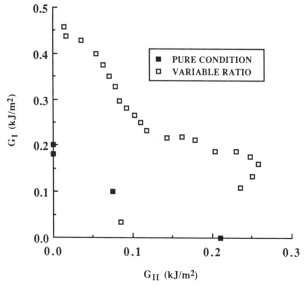

FIG. 9. Failure locus under mixed mode loading for bismaleimide.

9. For PEEK and epoxy there is a quite well defined locus for both forms of mixed mode test and there is little evidence of fibre bridging. For the varying ratio tests $1/L$ was varied to check independence of test method and the correctness of the analysis. The agreement between the two forms of test is encouraging. For bismaleimide the fibre bridging results in large discrepancies and the notion of a unique locus cannot be used. Since one would hope for good matrix-fibre adhesion in practice it is felt that the single line is likely to prove useful. The form of the curve is somewhat below a linear relationship between G_{IC} and G_{IIC}.

The quality of the data confirms that the test configurations and the analysis employed are satisfactory and represent a useful approach for analysing this form of fracture.

REFERENCES

1. WILLIAMS, J. G., *International Journal of Fracture*, **36** (1988) 101–19.
2. WILLIAMS, J. G., in: *Application of Fracture Mechanics to Composite Materials*, Chap. 1, (K. Friederich, ed.), to be published, Elsevier, 1989.
3. HASHEMI, S., KINLOCH, A. J. and WILLIAMS, J. G., *Proc. ICCM 6*, 3, 254.
4. HASHEMI, S., KINLOCH, A. J. and WILLIAMS, J. G., to be published.

5

Constitutive Relations for Transversely Isotropic Materials

G. F. Smith and G. Bao

*Center for the Application of Mathematics, Lehigh University,
Bethlehem, Pennsylvania, USA*

ABSTRACT

A procedure has been developed which enables one to employ a computer program to carry out the essential computations required for the generation of non-linear constitutive expressions for the cases where the material considered belongs to one of the 32 crystal classes. This procedure is outlined here and extended so as to also cover the cases where the group defining the material symmetry is one of the five transverse isotropy groups.

1. INTRODUCTION

Composite materials are replacing more traditional materials in a wide variety of applications. The objective is to design a material which has the properties required for the given application. We may take a first step in this direction if we can specify the form of the constitutive equation which would best suit out purposes. The form of a constitutive equation describing the response of a material is essentially dictated by the symmetry properties of the material. We may search through the various categories of constitutive equations associated with materials which possess symmetry properties in order to find which most closely matches the desired form. We may then hope to fabricate a material with the appropriate symmetry by judicious placement of fibers in a matrix comprised of an isotropic material. In order to aid in the recognition process, we write the constitutive expression in matrix form where the matrices describing the material properties are diagonalized to the extent possible. We employ methods from group representation theory to attain this end. We plan to consider non-linear constitutive expressions which generally require a substantial amount of computation. To overcome any difficulties in this direction, we have developed a computer program which will generate the diagonalized matrix form of a wide class of constitutive expressions for materials possessing the symmetry associated with any of the crystallographic groups.[1,2] In this study, we give the basic information required to extend these results to the cases where the material possesses symmetry properties characterized by one of the five transverse isotropy groups which we denote by $T_1, T_2, \ldots, T_5$. The group T_1 defines the symmetry of a material which possesses rotational symmetry about the x_3 axis. The group T_2 defines this symmetry of a material which possesses rotational symmetry about the x_3 axis and for which the plane containing the x_2 and x_3 axes is a plane of symmetry. The group T_3 defines the symmetry of a material possessing rotational symmetry about the x_3 axis and for which the plane containing the x_1 and x_2 axes is a plane of symmetry. The group T_4 is associated with a material possessing rotational symmetry about the x_3 axis and for which the planes containing the x_2 and x_3 axes and the x_1 and x_2 axes are both planes of symmetry. The group T_5 is associated with a material possessing rotational symmetry about the x_3 axis and for which the x_2 axis is a two-fold axis of symmetry.

We observe that the consideration of transversely isotropic materials may arise in connection with composites in the following ways. Consider a circular cylinder in which is embedded a number of fibers which are parallel to the axis of the cylinder. If the fibers are uniformly distributed over the cross-section of the cylinder, we may assume that the material possesses rotational symmetry about the axis of the cylinder. In addition, the plane perpendicular to the axis of the cylinder and the planes containing the axis of the cylinder are planes of symmetry. The group defining the material symmetry would be the transverse isotropy group denoted by T_4. One may fabricate a material consisting of layers of thin sheets in which an array of parallel fibers is present. If the fibers in any given layer make an angle of $2\pi/n$ with the fibers in the adjacent layers, we may consider the composite to have an n-fold axis of symmetry perpendicular to the layers. Further we may assume that the material possesses a plane of symmetry which contains the n-fold axis of symmetry. If n becomes large, this would render the material approximately transversely isotropic. The group defining the material symmetry would be the transverse isotropy group T_2.

We discuss below the procedure employed to generate the block diagonal form of the matrix constitutive equations. We then give the information required to employ this procedure for the transversely isotropic groups T_1, $T_2, \ldots, T_5$. We finally give examples of the application of these results to the generation of non-linear constitutive expressions. The work discussed below constitutes part of the doctoral dissertation[3] of one of the authors (G. Bao).

2. TRANSVERSELY ISOTROPIC MATERIALS

The constitutive expressions which we consider are tensor-valued functions of one or more tensors $\mathbf{S}_1, \mathbf{S}_2, \ldots$ of degrees $n_1, n_2, \ldots$ in these tensors which are invariant under a group Γ defining the material symmetry. We are primarily interested in the cases where the material symmetry is defined by one of the five groups $T_1, \ldots, T_5$ associated with the various types of transversely isotropic materials. These groups are defined by specifying the groups of 3×3 matrices which define the set of equivalent reference frames associated with the material. Thus the group T_1 is comprised of the matrices

$$\mathbf{Q}(\theta) = \left\| \begin{array}{ccc} \cos\theta, & \sin\theta, & 0 \\ -\sin\theta, & \cos\theta, & 0 \\ 0, & 0, & 1 \end{array} \right\|, \quad 0 \leqslant \theta < 2\pi \tag{1}$$

.Let $\mathbf{e}_i$ denote the base vectors associated with rectangular Cartesian coordinate system x. Let $\bar{\mathbf{e}}_i$ denote the base vectors associated with the reference frame $\bar{x}$ which is obtained by rotating the reference frame x through θ radians counter-clockwise. We have

$$\bar{\mathbf{e}}_i = Q_{ij}(\theta)\mathbf{e}_j \tag{2}$$

where the matrix $\mathbf{Q}(\theta) = \|Q_{ij}(\theta)\|$ is defined by (1). If x and $\bar{x}$ are equivalent reference frames, i.e. if $\bar{x}$ arises from x by applying a symmetry operation to x, the constitutive equation is required to have the same form when referred to either reference frame. Let T_{ij} and E_{ij} denote (absolute) second-order tensors. Then, if

$$T_{ij} = C_{ijkl\ldots mn} E_{kl} \ldots E_{mn} \tag{3}$$

is the constitutive expression when referred to the x frame, the constitutive expression where referred to the $\bar{x}$ frame is given by

$$\bar{T}_{ij} = \bar{C}_{ijkl\ldots mn} \bar{E}_{kl} \ldots \bar{E}_{mn} \tag{4}$$

where

$$\bar{T}_{ij} = Q_{ip}Q_{jq}T_{pq}, \quad \bar{C}_{ij...n} = Q_{ip}Q_{jq}\cdots Q_{nr}C_{pq...r}$$

$$\bar{E}_{kl} = Q_{ks}Q_{lt}E_{st} \tag{5}$$

If the reference frame x and $\bar{x}$ are equivalent, we require that

$$\bar{C}_{ijkl...mn} = C_{ijkl...mn} \tag{6}$$

If the symmetry operations consist of all rotations about the x_3 axis, the property tensor $C_{ijkl...mn}$ must satisfy the equations

$$Q_{ip}(\theta)Q_{jq}(\theta)\dots Q_{nu}(\theta)C_{pq...u} = C_{ij...n} \tag{7}$$

for $0 \leqslant \theta < 2\pi$. We say that the tensor $C_{ij...n}$ which satisfies eqn (7) for $0 \leqslant \theta < 2\pi$ is invariant under the group T_1. We may express a tensor which is invariant under the group T_1 as a linear combination of the outer products of the fundamental tensors

$$\delta_{3i}, \ \delta_{1i}\delta_{1j} + \delta_{2i}\delta_{2j} = \alpha_{ij}, \ \delta_{1i}\delta_{2j} - \delta_{2i}\delta_{1j} = \beta_{ij} \tag{8}$$

Thus, one may express the general fourth-order tensor which is invariant under the group T_1 as a linear combination of the

$$\begin{array}{llll}
\delta_{3i}\delta_{3j}\delta_{3k}\delta_{3l}, \ 1; & \alpha_{ij}\alpha_{kl}, \ 3; & \\
\delta_{3i}\delta_{3j}\alpha_{kl}, 6; & \alpha_{ij}\beta_{kl}, \ 6; & (9) \\
\delta_{3i}\delta_{3j}\beta_{kl}, 6; & \beta_{ij}\beta_{kl}, \ 3; &
\end{array}$$

where the numbers following the tensors denote the number of distinct isomers of the tensor. An isomer of a tensor σ_{ijkl} is obtained by permuting the subscripts i, j, k, l of the tensor. We have noted that $\alpha_{ij} = \alpha_{ji}$ and $\beta_{ij} = -\beta_{ji}$. We further note that

$$\beta_{ij}\beta_{kl} = \alpha_{ik}\alpha_{jl} - \alpha_{il}\alpha_{jk} \tag{10}$$

and that only three of the six isomers of $\alpha_{ij}\beta_{kl}$ are linearly independent. Thus, we have

$$\alpha_{ij}\beta_{kl} + \alpha_{ik}\beta_{lj} + \alpha_{il}\beta_{jk} = 0$$

$$\alpha_{ij}\beta_{kl} + \alpha_{kj}\beta_{li} + \alpha_{lj}\beta_{ik} = 0 \tag{11}$$

$$\alpha_{ik}\beta_{jl} + \alpha_{jk}\beta_{li} + \alpha_{lk}\beta_{ij} = 0$$

The existence of relations such as (11) renders the generation of the general tensors of orders 5, 6, ... which are invariant under T_1 a non-trivial matter. This may be accomplished using a method[4] which employs Young tableaux. However the procedure for generating the form of a constitutive equation based on listing the general form of the property tensor $C_{ijkl...mn}$ and then substituting into (3) would generally prove to be cumbersome. It is preferable to employ a procedure based on group representation theory. A set of matrices $\mathbf{P}(\theta)$ which is in one to one correspondence with the matrices $\mathbf{Q}(\theta)$ comprising the group T_1 and such that $\mathbf{P}(\theta_1)\mathbf{P}(\theta_2)$ corresponds to $\mathbf{Q}(\theta_1)\mathbf{Q}(\theta_2)$ is said to form a matrix representation of the group T_1. The set of matrices $\mathbf{K}\mathbf{P}(\theta)\mathbf{K}^{-1}$ where det $\mathbf{K} \neq 0$ also forms a matrix representation of T_1 which is said to be equivalent to the representation $\mathbf{P}(\theta)$. An appropriate choice of the matrix $\mathbf{K}$ enables us to write

$$\mathbf{KP}(\theta)\mathbf{K}^{-1} = \alpha_1 \mathbf{P}_1(\theta) \dotplus \alpha_2 \mathbf{P}_2(\theta) \dotplus \cdots \tag{12}$$

in block diagonal form where $\alpha_1, \alpha_2, \ldots$ are positive integers and where

$$2\mathbf{P}_1(\theta) \dotplus \mathbf{P}_2(\theta) = \begin{Vmatrix} \mathbf{P}_1(\theta), & 0, & 0 \\ 0, & \mathbf{P}_1(\theta), & 0 \\ 0, & 0, & \mathbf{P}_2(\theta) \end{Vmatrix} \tag{13}$$

We say that the representation $\mathbf{P}(\theta)$ may be decomposed into the direct sum of the representations $\mathbf{P}_1(\theta), \mathbf{P}_2(\theta), \ldots$. If a representation $\mathbf{P}(\theta)$ cannot be decomposed, it is referred to as an irreducible representation. The irreducible representations associated with T_1 are all one-dimensional and are defined by listing the 1×1 matrix corresponding to the matrix $\mathbf{Q}(\theta)$. We define these representations below.

$$\gamma_0 : 1$$
$$\gamma_p : e^{-ip\theta} \quad (p = 1, 2, \ldots) \tag{14}$$
$$\Gamma p : e^{ip\theta} \quad (p = 1, 2, \ldots)$$

γ_0 denotes the identity representation where the same number 1 corresponds to each $\mathbf{Q}(\theta)$; γ_p denote the representation where $e^{-ip\theta}$ corresponds to $\mathbf{Q}(\theta), \ldots$.

We consider the manner in which a vector transforms under the group T_1. The components $\bar{x}_i$ of a vector $\bar{\mathbf{x}}$ when referred to the reference frame $\bar{x}$ with base vectors $\bar{\mathbf{e}}_i = Q_{ij}(\theta)\mathbf{e}_j$ are related to the components x_i of $\mathbf{x}$ when referred to the x frame with base vectors $\mathbf{e}_i$ by the equations

$$\bar{x}_i = Q_{ij}(\theta)x_j \text{ or } \begin{Vmatrix} \bar{x}_1 \\ \bar{x}_2 \\ \bar{x}_3 \end{Vmatrix} = \begin{Vmatrix} \cos\theta, & \sin\theta, & 0 \\ -\sin\theta, & \cos\theta, & 0 \\ 0, & 0, & 1 \end{Vmatrix} \begin{Vmatrix} x_1 \\ x_2 \\ x_3 \end{Vmatrix} \tag{15}$$

With (15) we readily see that

$$\left\| \begin{array}{c} \bar{x}_1 + i\bar{x}_2 \\ \bar{x}_1 - i\bar{x}_2 \\ \bar{x}_3 \end{array} \right\| = \left\| \begin{array}{ccc} e^{-i\theta}, & 0, & 0 \\ 0, & e^{i\theta}, & 0 \\ 0, & 0, & 1 \end{array} \right\| \left\| \begin{array}{c} x_1 + ix_2 \\ x_1 - ix_2 \\ x_3 \end{array} \right\| \tag{16}$$

This tells us that the transformation properties of $x_1 + ix_2$, $x_1 - ix_2$, x_3 under the group T_1 are defined respectively by the irreducible representations γ_1, Γ_1 and γ_0 respectively. We immediately see that the transformation properties of

$$(x_1 + ix_2)^2, \ (x_1 + ix_2)(x_1 - ix_2), \ (x_1 + ix_2)x_3, \\ (x_1 - ix_2)^2, \ (x_1 - ix_2)x_3, \ x_3^2 \tag{17}$$

are defined by the irreducible representations γ_2, γ_0, γ_1, Γ_2, Γ_1, γ_0 respectively. Since the components $x_i x_j$ transform in the same manner as do the components S_{ij} of a symmetric second-order tensor, we see that the transformation properties of

$$S_{11} - S_{22} + 2iS_{12}, \ S_{11} + S_{22}, \ S_{13} + iS_{23}, \\ S_{11} - S_{22} - 2iS_{12}, \ S_{13} - iS_{23}, \ S_{33} \tag{18}$$

transform according to γ_2, γ_0, γ_1, Γ_2, Γ_1, γ_0 respectively. Similarly we see that

$$(S_{11} - S_{22} + 2iS_{12})^2, \ (S_{11} - S_{22} + 2iS_{12})(S_{11} + S_{22}), \\ (S_{11} - S_{22} + 2iS_{12})(S_{13} + iS_{23}), \\ (S_{11} - S_{22} + 2iS_{12})(S_{11} - S_{22} - 2iS_{12}), \dots \tag{19}$$

transform according to γ_4, γ_2, γ_3, γ_0, ... respectively. Thus, we may readily determine the linear combinations of the components of tensors $x_i, x_i x_j, x_i x_j x_k, \dots, S_{ij}, S_{ij} S_{kl}, \dots, S_{ij} x_k, S_{ij} x_k x_i, \dots$ which belong to the various irreducible representations $\gamma_0, \gamma_1, \gamma_2, \dots, \Gamma_1, \Gamma_2, \dots$ of the group Γ_1.

Let us consider the problem of determining the linear stress–strain relation for a material whose symmetry is defined by the group T_1. We write the constitutive expression as

$$\mathbf{T} = \mathbf{C}_1 \mathbf{E}_1 \tag{20}$$

where

$$\mathbf{T} = \| t_{11} + t_{22}, t_{33}, t_{11} - t_{22} + 2it_{12}, t_{11} - t_{22} - 2it_{12}, \\ t_{13} + it_{23}, t_{13} - it_{23} \|^{\mathrm{T}}$$

$$\mathbf{E}_1 = \|e_{11} + e_{22}, e_{33}, e_{11} - e_{22} + 2ie_{12}, e_{11} - e_{22} - 2ie_{12},$$

$$e_{13} + ie_{23}, e_{13} - ie_{23}\|^{\mathrm{T}} \tag{21}$$

and where $\mathbf{C}_1$ is a 6×6 matrix. If we refer the expression to the reference frame $\bar{x}$, whose base vectors are given by $\bar{\mathbf{e}}_i = Q_{ij}(\theta)\mathbf{e}_j$, we have

$$\bar{\mathbf{T}} = \bar{\mathbf{C}}_1\bar{\mathbf{E}}_1, \quad \bar{\mathbf{T}} = \mathbf{R}(\theta)\mathbf{T}, \quad \bar{\mathbf{E}}_1 = \mathbf{R}(\theta)\mathbf{E}_1, \quad \bar{\mathbf{C}}_1 = \mathbf{R}(\theta)\mathbf{C}_1 R^{-1}(\theta) \tag{22}$$

If the reference frame $\bar{x}$ is an equivalent reference frame, we require that $\bar{\mathbf{C}}_1 = \mathbf{C}_1$, i.e.

$$\mathbf{R}(\theta)\mathbf{C}_1 = \mathbf{C}_1\mathbf{R}(\theta) \tag{23}$$

We observe from (14) and (18) that

$$\mathbf{R}(\theta) = \mathrm{diag}\ (1,\ 1,\ e^{-2i\theta},\ e^{2i\theta},\ e^{-i\theta},\ e^{i\theta}) \tag{24}$$

With (23) and (24), we have 36 equations relating the entries $C_{ij}(i,j = 1, \ldots 6)$ of $\mathbf{C}_1$ which are given by

$$C_{11} = C_{11}, \quad C_{12} = C_{12}, \quad C_{13} = e^{-2i\theta}C_{13}, \quad C_{14} = e^{2i\theta}C_{14},$$
$$C_{15} = e^{-i\theta}C_{15}, \quad C_{16} = e^{i\theta}C_{16}, \ldots \tag{25}$$

With (25) we have

$$\mathbf{C}_1 = \begin{Vmatrix} C_{11}, & C_{12}, & 0, & 0, & 0, & 0 \\ C_{21}, & C_{22}, & 0, & 0, & 0, & 0 \\ 0, & 0, & C_{33}, & 0, & 0, & 0 \\ 0, & 0, & 0, & C_{44}, & 0, & 0 \\ 0, & 0, & 0, & 0, & C_{55}, & 0 \\ 0, & 0, & 0, & 0, & 0, & C_{66} \end{Vmatrix} \tag{26}$$

This tells us each entry in $\mathbf{T}$ which belongs to a representation γ_p is expressible as a linear combination of the elements of $\mathbf{E}_1$ which belong to γ_p. Thus, with (21) and (26), we see that $\mathbf{T} = \mathbf{C}_1\mathbf{E}_1$ may be written as

$$\begin{Vmatrix} t_{11} + t_{22} \\ t_{33} \end{Vmatrix} = \begin{Vmatrix} C_{11}, C_{12} \\ C_{21}, C_{22} \end{Vmatrix} \begin{Vmatrix} e_{11} + e_{22} \\ e_{33} \end{Vmatrix}, \gamma_0$$

$$t_{11} - t_{22} + 2it_{12} = C_{33}(e_{11} - e_{22} + 2ie_{12}),\ \gamma_2$$

$$t_{11} + t_{22} - 2it_{12} = C_{44}(e_{11} - e_{22} - 2ie_{12}),\ \Gamma_2 \tag{27}$$

$$t_{13} + it_{23} = C_{55}(e_{13} + ie_{23}),\ \gamma_1$$

$$t_{13} - it_{23} = C_{66}(e_{13} - ie_{23}),\ \Gamma_1$$

where the $\gamma_0, \ldots$ indicates that the quantities in the preceding equation belong to the irreducible representation $\gamma_0, \ldots$. In (27) it is clear that we should set $C_{44} = \bar{C}_{33}$ and $C_{66} = \bar{C}_{55}$. If we set $C_{33} = a + ib$, $C_{55} = c + id$, the expressions $(27)_3, \ldots, (27)_6$ may be written as

$$\left\| \begin{matrix} t_{11} - t_{22} \\ 2t_{12} \end{matrix} \right\| = \left\| \begin{matrix} a, -b \\ b, \; a \end{matrix} \right\| \left\| \begin{matrix} e_{11} - e_{22} \\ 2e_{12} \end{matrix} \right\|, \quad \left\| \begin{matrix} t_{13} \\ t_{23} \end{matrix} \right\| = \left\| \begin{matrix} c, -d \\ d, \; c \end{matrix} \right\| \left\| \begin{matrix} e_{13} \\ e_{23} \end{matrix} \right\| \tag{28}$$

Let us consider the case where the constitutive expression is given by

$$t_{ij} = c_{ijklmn} e_{kl} e_{mn}, \; t_{ij} = t_{ji}, \; e_{kl} = e_{lk} \tag{29}$$

We write this in matrix form as

$$\mathbf{T} = \mathbf{C}_2 \mathbf{E}_2 \tag{30}$$

where $\mathbf{T}$ is given by (21) and $\mathbf{E}_2$ denotes the (21×1) column matrix whose entries are linearly independent linear combinations of the 21 quantities $e_{11}^2, e_{11}e_{12}, \ldots$ so chosen that each belongs to one of the irreducible representations of T_1. With the notation

$$E_1 = e_{11} + e_{22}, \; E_2 = e_{33}, \; E_3 = e_{13} + ie_{23}, \; E_4 = e_{13} - ie_{23}$$
$$E_5 = e_{11} - e_{22} + 2ie_{12}, \; E_6 = e_{11} - e_{22} - 2ie_{12} \tag{31}$$

we find that the 21 quantities of degree 2 in the E_i which belong to $\gamma_0, \gamma_1, \Gamma_1, \ldots$ are given by

$$\gamma_0 : E_1^2, E_1 E_2, E_2^2, E_3 E_4, E_5 E_6;$$

$$\gamma_1 : E_1 E_3, E_2 E_3, E_4 E_5; \qquad \Gamma_1 : E_1 E_4, E_2 E_4, E_3 E_6;$$

$$\gamma_2 : E_1 E_5, E_2 E_5, E_3^2; \qquad \Gamma_2 : E_1 E_6, E_2 E_6, E_4^2; \tag{32}$$

$$\gamma_3 : E_3 E_5; \qquad \Gamma_3 : E_4 E_6;$$

$$\gamma_4 : E_5^2; \qquad \Gamma_4 : E_6^2$$

The constitutive expression $\mathbf{T} = \mathbf{C}_2 \mathbf{E}_2$ may then be written as

$$\left\| \begin{matrix} t_{11} + t_{22} \\ t_{33} \end{matrix} \right\| = \left\| \begin{matrix} c_1, c_2, c_3, c_4, c_5 \\ c_6, c_7, c_8, c_9, c_{10} \end{matrix} \right\| \left\| \begin{matrix} E_1^2 \\ E_1 E_2 \\ E_2^2 \\ E_3 E_4 \\ E_5 E_6 \end{matrix} \right\|, \gamma_0$$

$$t_{13} + it_{23} = c_{11}E_1E_3 + c_{12}E_2E_3 + c_{13}E_4E_5, \ \gamma_1$$

$$t_{13} - it_{23} = \bar{c}_{11}E_1E_4 + \bar{c}_{12}E_2E_4 + \bar{c}_{13}E_3E_6, \ \Gamma_1$$

$$t_{11} - t_{22} + 2it_{12} = c_{14}E_1E_5 + c_{15}E_2E_5 + c_{16}E_3^2, \ \gamma_2$$

$$t_{11} - t_{22} - 2it_{12} = \bar{c}_{14}E_1E_6 + \bar{c}_{15}E_2E_6 + \bar{c}_{16}E_4^2, \ \Gamma_2 \tag{33}$$

where we have noted that $E_4 = \bar{E}_3$ and $E_6 = \bar{E}_5$.

3. THE IRREDUCIBLE REPRESENTATIONS FOR THE TRANSVERSELY ISOTROPIC GROUPS

There are five groups which we refer to as transversely isotropic and which are denoted by $T_1, \ldots, T_5$. These groups are defined by listing matrices such that these matrices or products of these matrices specify all of the symmetry operations associated with the material under consideration. We may refer to these matrices as generators of the group. We then define the irreducible representations associated with a group T_i by listing the matrices which correspond to the generators of the group.

Suppose that we are given that the quantities $a_1, a_2, \|a_3, a_4\|^T, \|a_5, a_6\|^T, \ldots$ belong to the irreducible representations $\gamma_0, \gamma_1, \gamma_2, \gamma_3, \ldots$ of a group and that the quantities $b_1, b_2, \|b_3, b_4\|^T, \ \|b_5, b_6\|^T, \ldots$ belong to the irreducible representations $\gamma_0, \gamma_1, \gamma_2, \gamma_3, \ldots$ of the same group. We need to determine the linear combinations $c_i = a_{ijk}a_jb_k$ of the products of the a_j and b_k which belong to the various irreducible representations of the group. This information is provided below for the groups $T_1, \ldots, T_5$ in tables which are referred to as product tables. Xu *et al.*[2] have indicated how these tables may be employed in conjunction with a computer program to automatically generate constitutive expressions. The extension of the results in Ref. 2 to include the transversely isotropic materials requires the development of a number of computer programs. This work will be carried out subsequently.

We further list in tables entitled basic quantities the linear combinations of the components of polar vectors p_i, axial vectors a_i and symmetric second-order tensors S_{ij} which belong to the various irreducible representations of the group considered.

3.1. The Group T_1

The group T_1 is comprised of the matrices $\mathbf{Q}(\theta)$ defined by

$$\mathbf{Q}(\theta) = \left\| \begin{array}{ccc} \cos\theta, & \sin\theta, & 0 \\ -\sin\theta, & \cos\theta, & 0 \\ 0, & 0, & 1 \end{array} \right\|, \ 0 \leqslant \theta < 2\pi \tag{34}$$

The group T_1 defines the symmetry of a material which possesses rotational symmetry about the x_3 axis. The irreducible representations associated with the group T_1 are all one-dimensional and are given[5] by

$$\begin{aligned} \gamma_0: & \ 1 \\ \gamma_p: & \ e^{-ip\theta} \ (p = 1,2,\ldots.) \\ \Gamma_p: & \ e^{ip\theta} \ (p = 1,2,\ldots.) \end{aligned} \tag{35}$$

In (35) the 1×1 matrices 1, $e^{-ip\theta}$ and $e^{ip\theta}$ correspond to the group element $\mathbf{Q}(\theta)$. The product table is given in Table 1, with the basic quantities in Table 2.

3.2. The Group T_2

The group T_2 is comprised of the matrices $\mathbf{Q}(\theta)$ and $\mathbf{R}_1\mathbf{Q}(\theta)$ where

TABLE 1
Product table for T_1

γ_0:	a_0, b_0
	$a_0 b_0$
	$a_p B_p, A_p b_p \ (p = 1,2,\ldots.)$
γ_p:	a_p, b_p
	$a_0 b_p, b_0 a_p$
	$a_m b_n \ (m,n, = 1,2,\ldots.; \ m+n=p)$
	$a_m B_n, A_n b_m \ (m,n = 1,2,\ldots; \ m-n=p)$
Γ_p:	A_p, B_p
	$a_0 B_p, A_p b_0$
	$A_m B_n \ (m,n = 1,2,\ldots; \ m+n=p)$
	$A_m b_n, a_n B_m \ (m,n = 1,2,\ldots; \ m-n=p)$

TABLE 2
Basic quantities for T_1

γ_0:	$p_3, a_3, S_{11}+S_{22}, S_{33}$
γ_1:	$p_1+ip_2, a_1+ia_2, S_{13}+iS_{23}$
Γ_1:	$p_1-ip_2, a_1-ia_2, S_{13}-iS_{23}$
γ_2:	$S_{11}-S_{22}+2iS_{12}$
Γ_2:	$S_{11}-S_{22}-2iS_{12}$

$0 \leqslant \theta \leqslant 2\pi$ and where

$$\mathbf{Q}(\theta) = \left\| \begin{array}{ccc} \cos\theta, & \sin\theta, & 0 \\ -\sin\theta, & \cos\theta, & 0 \\ 0, & 0, & 1 \end{array} \right\|, \mathbf{R}_1 = \mathrm{diag}(-1,1,1) \tag{36}$$

The irreducible representations associated with the group T_2 are defined by listing the matrices corresponding to the group elements $\mathbf{Q}(\theta)$ and $\mathbf{R}_1$. We denote the irreducible representations (see Ref. 5) by

$$\gamma_0: \quad 1, \; 1$$
$$\Gamma_0: \quad 1, \, -1$$
$$\gamma_p: \left\| \begin{array}{cc} e^{-ip\theta}, & 0 \\ 0, & e^{ip\theta} \end{array} \right\|, \left\| \begin{array}{cc} 0, & 1 \\ 1, & 0 \end{array} \right\| (p=1,2,\ldots) \tag{37}$$

TABLE 3
Product table for T_2

γ_0:	$a_0, b_0;$
	$a_0 b_0, A_0 B_0$
	$a_{m1}b_{m2}+a_{m2}b_{m1}$ $(m=1,2,\ldots.)$
Γ_0:	A_0, B_0
	$a_0 B_0, A_0 b_0$
	$a_{m1}b_{m2}-a_{m2}b_{m1}$ $(m=1,2,\ldots.)$
γ_p:	$\|a_{p1}, a_{p2}\|^T, \|b_{p1}, b_{p2}\|^T$
	$\|a_0 b_{p1}, a_0 b_{p2}\|^T, \|a_{p1}b_0, a_{p2}b_0\|^T$
	$\|A_0 b_{p1}, -A_0 b_{p2}\|^T, \|a_{p1}B_0, -a_{p2}B_0\|^T$
	$\|a_{m1}b_{n1}, a_{m2}b_{n2}\|^T, (m,n=1,2,\ldots.; m+n=p),$
	$\|a_{m1}b_{n2}, a_{m2}b_{n1}\|^T, \|a_{n2}b_{m1}, a_{n1}b_{m2}\|^T, (m,n=1,2,\ldots.; m-n=p)$

TABLE 4

Basic quantities for T_2

γ_0:	$p_3, S_{11} + S_{22}, S_{33}$
Γ_0:	a_3
γ_1:	$\|p_1 + ip_2, -p_1 + ip_2\|^T, \|a_1 + ia_2, a_1 - ia_2\|^T,$
	$\|S_{13} + iS_{23}, -S_{13} + iS_{23}\|^T$
γ_2:	$\|S_{11} - S_{22} + 2iS_{12}, S_{11} - S_{22} - 2iS_{12}\|^T$

where the first and second matrices correspond to $\mathbf{Q}(\theta)$ and $\mathbf{R}_1$ respectively. The product table is given in Table 3 and basic quantities in Table 4.

3.3. The Group T_3

The group T_3 is comprised of the matrices $\mathbf{Q}(\theta)$ and $\mathbf{R}_3\mathbf{Q}(\theta)$ where $0 \leqslant \theta < 2\pi$ and where

$$\mathbf{Q}(\theta) = \left\|\begin{array}{ccc} \cos\theta, & \sin\theta, & 0 \\ -\sin\theta, & \cos\theta, & 0 \\ 0, & 0, & 1 \end{array}\right\|, \quad \mathbf{R}_3 = \text{diag}(1,1,-1) \tag{38}$$

The irreducible representations associated with the group T_3 are defined by listing the matrices corresponding to the group elements $\mathbf{Q}(\theta)$ and $\mathbf{R}_3$. We denote the irreducible representations by

$$\begin{aligned}
&\gamma_0: \quad 1, 1 \\
&\Gamma_0: \quad 1, -1 \\
&\gamma_p: \quad e^{-ip\theta}, 1; \bar{\gamma}_p: e^{ip\theta}, 1 \ (p = 1, 2, \ldots.) \\
&\Gamma_p: \quad e^{-ip\theta}, -1; \bar{\Gamma}_p: e^{ip\theta}, -1 \ (p = 1, 2, \ldots.)
\end{aligned} \tag{39}$$

where the first and second 1×1 matrices correspond to $\mathbf{Q}(\theta)$ and $\mathbf{R}_3$ respectively. The product table is given in Table 5 and basic quantities in Table 6.

3.4. The Group T_4

The group T_4 is comprised of the matrices $\mathbf{Q}(\theta)$, $\mathbf{R}_1\mathbf{Q}(\theta)$, $\mathbf{R}_3\mathbf{Q}(\theta)$ and $\mathbf{R}_1\mathbf{R}_3\mathbf{Q}(\theta)$ where $0 \leqslant \theta < 2\pi$ and where

$$\mathbf{Q}(\theta) = \left\|\begin{array}{ccc} \cos\theta, & \sin\theta, & 0 \\ -\sin\theta, & \cos\theta & 0 \\ 0, & 0, & 1 \end{array}\right\|, \quad \mathbf{R}_1 = \text{diag}(-1,1,1), \mathbf{R}_3 = \text{diag}(1,1,-1) \tag{40}$$

TABLE 5
Product table, T_3

γ_0:	a_0, b_0
	$a_0 b_0, A_0 B_0$
	$a_m \bar{b}_m, \bar{a}_m b_m, A_m \bar{B}_m, \bar{A}_m B_m \ (m = 1, 2, \ldots)$
Γ_0:	A_0, B_0
	$a_0 B_0, A_0 b_0$
	$a_m \bar{B}_m, \bar{a}_m B_m, A_m \bar{b}_m, \bar{A}_m b_m \ (m = 1, 2, \ldots)$
γ_p:	a_p, b_p
	$a_0 b_p, a_p b_0, A_0 B_p, A_p B_0$
	$a_m b_n, A_m B_n \ (m, n = 1, 2, \ldots; m + n = p)$
	$a_m \bar{b}_n, \bar{a}_n b_m, A_m \bar{B}_n, \bar{A}_n B_m \ (m, n = 1, 2, \ldots; m - n = p)$
$\bar{\gamma}_p$:	$\bar{a}_p, \bar{b}_p$
	$a_0 \bar{b}_p, \bar{a}_p b_0, A_0 \bar{B}_p, \bar{A}_p B_0$
	$\bar{a}_m \bar{b}_n, \bar{A}_m \bar{B}_n \ (m, n = 1, 2, \ldots; m + n = p)$
	$\bar{a}_m b_n, a_n \bar{b}_m, \bar{A}_m B_n, A_n \bar{B}_m \ (m, n = 1, 2, \ldots; m - n = p)$
Γ_p:	$A_p, B_p;$
	$a_0 B_p, A_p b_0, A_0 b_p, a_p B_0$
	$a_m B_n, A_m b_n \ (m, n = 1, 2, \ldots; m + n = p)$
	$a_m \bar{B}_n, \bar{a}_n B_m, A_m \bar{b}_n, \bar{A}_n b_m \ (m, n = 1, 2, \ldots; m - n = p)$
$\bar{\Gamma}_p$:	$\bar{A}_p, \bar{B}_p$
	$a_0 \bar{B}_p, \bar{A}_p b_0, A_0 \bar{b}_p, \bar{a}_p B_0$
	$\bar{a}_m \bar{B}_n, \bar{A}_m \bar{b}_n, (m, n = 1, 2, \ldots; m + n = p)$
	$\bar{a}_m B_n, a_n \bar{B}_m, \bar{A}_m b_n, A_n \bar{b}_m \ (m, n = 1, 2, \ldots; m - n = p)$

TABLE 6
Basic quantities, T_3

γ_0:	$a_3, S_{11} + S_{22}, S_{33}$
Γ_0:	p_3
γ_1:	$p_1 + i p_2$
$\bar{\gamma}_1$:	$p_1 - i p_2$
Γ_1:	$a_1 + i a_2, S_{13} + i S_{23}$
$\bar{\Gamma}_1$:	$a_1 - i a_2, S_{13} - i S_{23}$
γ_2:	$S_{11} - S_{22} + 2 i S_{12}$
$\bar{\gamma}_2$:	$S_{11} - S_{22} - 2 i S_{12}$

The irreducible representations associated with the group T_4 are defined by listing the matrices corresponding to the group elements $Q(\theta)$, R_1 and R_3. We denote the irreducible representations by

$$\gamma_{01}: \quad 1, \quad 1, \quad 1$$
$$\gamma_{02}: \quad 1, \quad 1, -1$$
$$\gamma_{03}: \quad 1, -1, \quad 1$$
$$\gamma_{04}: \quad 1, -1, -1 \tag{41}$$

$$\gamma_p: \quad \left\|\begin{matrix} e^{-ip\theta}, & 0 \\ 0, & e^{ip\theta} \end{matrix}\right\|, \quad \left\|\begin{matrix} 0, & 1 \\ 1, & 0 \end{matrix}\right\|, \quad \left\|\begin{matrix} 1, & 0 \\ 0, & 1 \end{matrix}\right\| \quad (p=1,2,\ldots.)$$

$$\Gamma_p: \quad \left\|\begin{matrix} e^{-ip\theta}, & 0 \\ 0, & e^{ip\theta} \end{matrix}\right\|, \quad \left\|\begin{matrix} 0, & 1 \\ 1, & 0 \end{matrix}\right\|, \quad \left\|\begin{matrix} -1, & 0 \\ 0, & -1 \end{matrix}\right\| \quad (p=1,2,\ldots)$$

where the first, second and third matrices correspond to $Q(\theta)$, R_1 and R_3 respectively. The product table is shown in Table 7 and basic quantities in Table 8.

3.5. The Group T_5

The group T_5 is comprised of the matrices $Q(\theta)$ and $D_2 Q(\theta)$ where $0 \leqslant \theta < 2\pi$ and where

$$Q(\theta) = \left\|\begin{matrix} \cos\theta, & \sin\theta, & 0 \\ -\sin\theta, & \cos\theta, & 0 \\ 0, & 0, & 1 \end{matrix}\right\|, \quad D_2 = \text{diag}(-1, \quad 1, \quad -1) \tag{42}$$

The irreducible representations associated with the group T_5 are defined by listing the matrices corresponding to the group elements $Q(\theta)$ and D_2. We denote the irreducible representations by

$$\gamma_0: \quad 1, \quad 1$$
$$\Gamma_0: \quad 1, \quad -1$$
$$\gamma_p: \quad \left\|\begin{matrix} e^{-ip\theta}, & 0 \\ 0, & e^{ip\theta} \end{matrix}\right\|, \quad \left\|\begin{matrix} 0, & 1 \\ 1, & 0 \end{matrix}\right\| \quad (p=1,2,\ldots.) \tag{43}$$

TABLE 7

Product table for T_4

γ_{01}: a_1, b_1

  $a_1b_1, a_2b_2, a_3b_3, a_4b_4$

  $a_{m1}b_{m2} + a_{m2}b_{m1}, A_{m1}B_{m2} + A_{m2}B_{m1}$ $(m = 1, 2, \ldots)$

γ_{02}: a_2, b_2

  $a_1b_2, a_2b_1, a_3b_4, a_4b_3$

  $a_{m1}B_{m2} + a_{m2}B_{m1}, A_{m1}b_{m2} + A_{m2}b_{m1}$ $(m = 1, 2, \ldots)$

γ_{03}: a_3, b_3

  $a_1b_3, a_2b_4, a_3b_1, a_4b_2$

  $a_{m1}b_{m2} - a_{m2}b_{m1}, A_{m1}B_{m2} - A_{m2}B_{m1}$ $(m = 1, 2, \ldots)$

γ_{04}: a_4, b_4

  $a_1b_4, a_2b_3, a_3b_2, a_4b_1$

  $a_{m1}B_{m2} - a_{m2}B_{m1}, A_{m1}b_{m2} - A_{m2}b_{m1}$ $(m = 1, 2, \ldots)$

γ_p: $\|a_{p1}, a_{p2}\|^T, \|b_{p1}, b_{p2}\|^T$

  $\|a_1b_{p1}, a_1b_{p2}\|^T, \|a_{p1}b_1, a_{p2}b_1\|^T$

  $\|a_2B_{p1}, a_2B_{p2}\|^T, \|A_{p1}b_2, A_{p2}b_2\|^T$

  $\|a_3b_{p1}, -a_3b_{p2}\|^T, \|a_{p1}b_3, -a_{p2}b_3\|^T$

  $\|a_4B_{p1}, -a_4B_{p2}\|^T, \|A_{p1}b_4, -A_{p2}b_4\|^T$

  $\|a_{m1}b_{n1}, a_{m2}b_{n2}\|^T, \|A_{m1}B_{n1}, A_{m2}B_{n2}\|^T$ $(m, n = 1, 2, \ldots; m+n = p)$

  $\|a_{m1}b_{n2}, a_{m2}b_{n1}\|^T, \|a_{n2}b_{m1}, a_{n1}b_{m2}\|^T$ $(m, n = 1, 2, \ldots; m-n = p)$

  $\|A_{m1}B_{n2}, A_{m2}B_{n1}\|^T, \|A_{n2}B_{m1}, A_{n1}B_{m2}\|^T$ $(m, n = 1, 2, \ldots; m-n = p)$

Γ_p: $\|A_{p1}, A_{p2}\|^T, \|B_{p1}, B_{p2}\|^T$

  $\|a_1B_{p1}, a_1B_{p2}\|^T, \|A_{p1}b_1, A_{p2}b_1\|^T$

  $\|a_2b_{p1}, a_2b_{p2}\|^T, \|a_{p1}b_2, a_{p2}b_2\|^T$

  $\|a_3B_{p1}, -a_3B_{p2}\|^T, \|A_{p1}b_3, -A_{p2}b_3\|^T$

  $\|a_4b_{p1}, -a_4b_{p2}\|^T, \|a_{p1}b_4, -a_{p2}b_4\|^T$

  $\|a_{m1}B_{n1}, a_{m2}B_{n2}\|^T, \|A_{m1}b_{n1}, A_{m2}b_{n2}\|^T$ $(m, n = 1, 2, \ldots; m+n = p)$

  $\|a_{m1}B_{n2}, a_{m2}B_{n1}\|^T, \|a_{n2}B_{m1}, a_{n1}B_{m2}\|^T$ $(m, n = 1, 2, \ldots; m-n = p)$

  $\|A_{m1}b_{n2}, A_{m2}b_{n1}\|^T, \|A_{n2}b_{m1}, A_{n1}b_{m2}\|^T$ $(m, n = 1, 2, \ldots; m-n = p)$

---

where the first and second matrices correspond to $\mathbf{Q}(\theta)$ and $\mathbf{D}_2$ respectively. We note that the irreducible representations (43) are the same as those appearing in (37). The product table will then be the same as the product table for T_2 which is shown in Table 9, with basic quantities in Table 10.

TABLE 8

Basic quantities for T_4

γ_{01}:	$S_{11} + S_{22}, S_{33}$
γ_{02}:	p_3
γ_{03}:	a_3
γ_{04}:	
γ_1:	$\|p_1 + ip_2, -p_1 + ip_2\|^T$
Γ_1:	$\|a_1 + ia_2, a_1 - ia_2\|^T, \|S_{13} + iS_{23}, -S_{13} + iS_{23}\|^T$
γ_2:	$\|S_{11} - S_{22} + 2iS_{12}, S_{11} - S_{22} - 2iS_{12}\|^T$

TABLE 9

Product table for T_5

γ_p:	a_0, b_0
	$a_0 b_0, A_0 B_0$
	$a_{m1} b_{m2} + a_{m2} b_{m1} \quad (m = 1, 2, \ldots)$
Γ_0:	A_0, B_0
	$a_0 B_0, A_0 b_0$
	$a_{m1} b_{m2} - a_{m2} b_{m1} \quad (m = 1, 2, \ldots.)$
γ_p:	$\|a_{p1}, a_{p2}\|^T, \|b_{p1}, b_{p2}\|^T$
	$\|A_0 b_{p1}, -A_0 b_{p2}\|^T, \|a_{p1} B_0, -a_{p2} B_0\|^T$
	$\|a_0 b_{p1}, a_0 b_{p2}\|^T, \|a_{p1} b_0, a_{p2} b_0\|^T$
	$\|a_{m1} b_{n1}, a_{m2} b_{n2}\|^T, (m, n = 1, 2, \ldots.; m + n = p)$
	$\|a_{m1} b_{n2}, a_{m2} b_{n1}\|^T, \|a_{n2} b_{m1}, a_{n1} b_{m2}\|^T \quad (m, n = 1, 2, \ldots.; m - n = p)$

TABLE 10

Basic quantities, T_5

γ_0:	$S_{11} + S_{22}, S_{33}$
Γ_0:	p_3, a_3
γ_1:	$\|p_1 + ip_2, -p_1 + ip_2\|^T, \|a_1 + ia_2, -a_1 + ia_2\|^T, \|S_{13} + iS_{23}, S_{13} - iS_{23}\|^T$
γ_2:	$\|S_{11} - S_{22} + 2iS_{12}, S_{11} - S_{22} - 2iS_{12}\|^T$

4. APPLICATIONS

In this section, we give some examples of the application of the results derived above to the generation of non-linear constitutive expressions. We first consider the problem of determining the form of a second-order tensor-valued function

$$T_{ij} = C_{ijklm} x_k x_l x_m \tag{44}$$

which is of degree three in the components of a polar vector x_i and which is invariant under the group T_1. From Table 2 we see that

$$t_{11} + t_{22}, t_{33}, t_{13} + it_{23}, t_{13} - it_{23}, t_{11} - t_{22} + 2it_{12}, t_{11} - t_{22} - 2it_{12} \tag{45}$$

belong to $\gamma_0, \gamma_0, \gamma_1, \Gamma_1, \gamma_2, \Gamma_2$ respectively and that

$$x_3, x_1 + ix_2, x_1 - ix_2 \tag{46}$$

belong to $\gamma_0, \gamma_1, \Gamma_1$ respectively. Upon employing Table 1 twice, we see that

$$x_3^3, x_3(x_1^2 + x_2^2), x_3^2(x_1 + ix_2), (x_1^2 + x_2^2)(x_1 + ix_2), x_3^2(x_1 - ix_2), (x_1^2 + x_2^2)(x_1 - ix_2),$$

$$x_3(x_1^2 - x_2^2 + 2ix_1 x_2), x_3(x_1^2 - x_2^2 - 2ix_1 x_2), (x_1 + ix_2)^3, (x_1 - ix_2)^3 \tag{47}$$

belong to $\gamma_0, \gamma_0, \gamma_1, \gamma_1, \Gamma_1, \Gamma_1, \gamma_2, \Gamma_2, \gamma_3, \Gamma_3$ respectively. Each of the quantities in (45) which belongs to a representation γ_p (say) is expressible as a linear combination of the quantities in (46) which belong to γ_p. Thus, we have

$$\left\| \begin{matrix} t_{11} + t_{22} \\ \\ t_{33} \end{matrix} \right\| = \left\| \begin{matrix} c_1, c_2 \\ \\ c_3, c_4 \end{matrix} \right\| \left\| \begin{matrix} x_3^3 \\ \\ x_3(x_1^2 + x_2^2) \end{matrix} \right\|$$

$$t_{13} + it_{23} = c_5 x_3^2(x_1 + ix_2) + c_6(x_1^2 + x_2^2)(x_1 + ix_2)$$

$$t_{13} - it_{23} = \bar{c}_5 x_3^2(x_1 - ix_2) + \bar{c}_6(x_1^2 + x_2^2)(x_1 - ix_2) \tag{48}$$

$$t_{11} - t_{22} + 2it_{12} = c_7 x_3(x_1^2 - x_2^2 + 2ix_1 x_2)$$

$$t_{11} - t_{22} - 2it_{12} = \bar{c}_7 x_3(x_1^2 - x_2^2 - 2ix_1 x_2)$$

where $\bar{c}_5, \ldots, \bar{c}_7$ denote the complex conjugates of $c_5, \ldots, c_7$.

We next consider the problem of determining the form of the polar vector-valued function

$$x_i = c_{ijk} S_{jk} + c_{ijklm} S_{jk} S_{lm} \tag{49}$$

of the symmetric second-order tensor S_{ij} which is invariant under the group

T_2. We employ the notation

$$S_1 = S_{11} + S_{22}, S_2 = S_{33}, S_3 = S_{13} + iS_{23}, S_4 = -S_{13} + iS_{23}$$
$$S_5 = S_{11} - S_{22} + 2iS_{12}, S_6 = S_{11} - S_{22} - 2iS_{12} \tag{50}$$

We note that $S_4 = -\bar{S}_3$ and $S_6 = \bar{S}_5$. From Table 4 we see that

$$x_3, \|x_1 + ix_2, -x_1 + ix_2\|^T \tag{51}$$

belong to γ_0 and γ_1 respectively and that

$$S_1, S_2, \|S_3, S_4\|^T, \|S_5, S_6\|^T \tag{52}$$

belong to $\gamma_0, \gamma_0, \gamma_1$ and γ_2 respectively. In (52) we have used the notation (50). Upon employing Table 3, we see that the 21 products of degree two in the $S_i (i = 1, \ldots, 6)$ belong to the representations listed below.

$$\gamma_0: \quad S_1^2, S_2^2, S_1 S_2, S_3 S_4, S_5 S_6$$
$$\gamma_1: \quad S_1 \|S_3, S_4\|^T, S_2 \|S_3, S_4\|^T, \|S_4 S_5, S_3 S_6\|^T$$
$$\gamma_2: \quad S_1 \|S_5, S_6\|^T, S_2 \|S_5, S_6\|^T, \|S_3^2, S_4^2\|^T \tag{53}$$
$$\gamma_3: \quad \|S_3 S_5, S_4 S_6\|^T$$
$$\gamma_4: \quad \|S_5^2, S_6^2\|^T$$

The constitutive equation (49) may then be written as

$$x_3 = c_1 S_1 + c_2 S_2 + c_3 S_1^2 + c_4 S_2^2 + c_5 S_1 S_2 + c_6 S_3 S_4 + c_7 S_5 S_6$$

$$\left\| \begin{matrix} x_1 + ix_2 \\ -x_1 + ix_2 \end{matrix} \right\| = \left\| \begin{matrix} c_8, 0 \\ 0, c_8 \end{matrix} \right\| \left\| \begin{matrix} S_3 \\ S_4 \end{matrix} \right\| + \left\| \begin{matrix} c_9, 0 \\ 0, c_9 \end{matrix} \right\| \left\| \begin{matrix} S_1 S_3 \\ S_1 S_4 \end{matrix} \right\| + \left\| \begin{matrix} c_{10}, 0 \\ 0, c_{10} \end{matrix} \right\| \left\| \begin{matrix} S_2 S_3 \\ S_2 S_4 \end{matrix} \right\|$$

$$+ \left\| \begin{matrix} c_{11}, 0 \\ 0, c_{11} \end{matrix} \right\| \left\| \begin{matrix} S_4 S_5 \\ S_3 S_6 \end{matrix} \right\| \tag{54}$$

Since $-x_1 + ix_2 = -\overline{(x_1 + ix_2)}$ and $S_4 = -\bar{S}_3$, we see that $c_8 = \bar{c}_8$ which implies that c_8 is a real number. Similarly, we see that c_9, c_{10} and c_{11} are also real numbers.

We next consider the problem of determining the form of the polar vector-valued function

$$x_i = c_{ijk} S_{jk} + c_{ijklm} S_{jk} S_{lm} \tag{55}$$

and the axial vector-valued function

$$a_i = d_{ijk} S_{jk} + d_{ijklm} S_{jk} S_{lm} \tag{56}$$

of the symmetric second-order tensor S_{ij} which is invariant under the group T_3. We employ the notation

$$S_1 = S_{11} + S_{22}, S_2 = S_{33}, S_3 = S_{13} + iS_{23}, S_4 = S_{13} - iS_{23}$$
$$S_5 = S_{11} - S_{22} + 2iS_{12}, S_6 = S_{11} - S_{11} - S_{22} - 2iS_{12} \tag{57}$$

We note that $S_4 = \bar{S}_3$ and $S_6 = \bar{S}_5$. From Table 6 we see that

$$x_3, x_1 + ix_2, x_1 - ix_2 \tag{58}$$

belong to $\Gamma_0, \gamma_1, \bar{\gamma}_1$ respectively, that

$$a_3, a_1 + ia_2, a_1 - ia_2 \tag{59}$$

belong to $\gamma_0, \Gamma_1, \bar{\Gamma}_1$ respectively and that

$$S_1, S_2, S_3, S_4, S_5, S_6 \tag{60}$$

belong to $\gamma_0, \gamma_0, \Gamma_1, \bar{\Gamma}_1, \gamma_2, \bar{\gamma}_2$ respectively. Upon employing Table 5, we see that the 21 products of degree 2 in the $S_i(i=1,\ldots,6)$ belong to the representations listed below.

$$\gamma_0: \quad S_1^2, S_2^2, S_1 S_2, S_3 S_4, S_5 S_6$$

$$\Gamma_1: \quad S_1 S_3, S_2 S_3, S_4 S_5$$

$$\bar{\Gamma}_1: \quad S_1 S_4, S_2 S_4, S_3 S_6$$

$$\gamma_2: \quad S_1 S_5, S_2 S_5, S_3^2$$

$$\bar{\gamma}_2: \quad S_1 S_6, S_2 S_6, S_4^2 \tag{61}$$

$$\Gamma_3: \quad S_3 S_5$$

$$\bar{\Gamma}_3: \quad S_4 S_6$$

$$\gamma_4: \quad S_5^2$$

$$\bar{\gamma}_4: \quad S_6^2$$

With (58), ... (61), we see that there are no terms of degrees 1 or 2 in the $S_i(i=1,\ldots,6)$ which belong to any of the representations to which $x_3, x_1 + ix_2, x_1 - ix_2$ belong. This implies that a constitutive expression of the form (55) is ruled out by symmetry considerations. The constitutive expression (56) may be written as

$$a_3 = d_1 S_1 + d_2 S_2 + d_3 S_1^2 + d_4 S_2^2 + d_5 S_1 S_2 + d_6 S_3 S_4 + d_7 S_5 S_6$$
$$a_1 + ia_2 = d_8 S_3 + d_9 S_1 S_3 + d_{10} S_2 S_3 + d_{11} S_4 S_5 \tag{62}$$
$$a_1 - ia_2 = \bar{d}_8 S_4 + \bar{d}_9 S_1 S_4 + \bar{d}_{10} S_2 S_4 + \bar{d}_{11} S_3 S_6$$

With the notation $d_8 = e_8 + if_8, \ldots, d_{11} = e_{11} + if_{11}$, we see from (57) that $(62)_{2,3}$ may be written as

$$\left\Vert\begin{matrix} a_1 \\ a_2 \end{matrix}\right\Vert = \left\Vert\begin{matrix} e_8, & -f_8 \\ f_8, & e_8 \end{matrix}\right\Vert \left\Vert\begin{matrix} S_{13} \\ S_{23} \end{matrix}\right\Vert + \cdots + \left\Vert\begin{matrix} e_{11}, & -f_{11} \\ f_{11}, & e_{11} \end{matrix}\right\Vert \left\Vert\begin{matrix} (S_{11} - S_{22})S_{13} + 2S_{12}S_{23} \\ -(S_{11} - S_{22})S_{23} + 2S_{12}S_{13} \end{matrix}\right\Vert$$

(63)

REFERENCES

1. SMITH, G. F., and KIRAL, E., Anisotropic constitutive equations and Schur's lemma, *Int. J. Engng Sci.*, **16** (1978) 773–80.
2. XU, Y. H., SMITH, M. M. and SMITH, G. F., Computer aided generation of anisotropic constitutive equations, *Int. J. Engng Sci.*, **25** (1987) 711–22.
3. BAO, G., Application of group and invariant-theoretic methods to the generation of constitutive equations, Ph.D. Dissertation, Lehigh University, Bethlehem, Pa., 1987.
4. SMITH G. F., On isotropic tensors and rotation tensors of dimension *m* and order *n*, *Tensor, N.S.*, **19** (1968) 79–88.
5. VAN DER WAERDEN, B. L., *Group Theory and Quantum Mechanics*, New York, Heidelberg, Berlin, Springer-Verlag, 1974.

6

Transverse Matrix Cracking in Composite Laminates

N. LAWS

Department of Mechanical Engineering, University of Pittsburgh,
Pittsburgh, Pennsylvania, USA

and

G. J. DVORAK

Department of Civil Engineering, Rensselaer Polytechnic Institute,
Troy, New York, USA

ABSTRACT

This paper focusses on the study of transverse cracking in cross-ply composite laminates under monotonic load. In particular an explicit formula is obtained for the loss of stiffness of a cracked laminate. But the major part of the paper is devoted to the statistical fracture mechanics of progressive cracking. We show how to predict crack density as a function of applied load. The model agrees well with experiment.

1. INTRODUCTION

Problems associated with transverse matrix cracking in cross-ply composite laminates have been discussed by many investigators. Two particular goals have been to predict loss of stiffness and crack density under monotonic tensile load. To the best of our knowledge the problem was first addressed by Garrett and Bailey.[1] The model proposed by these authors is usually referred to as shear lag theory. This model was subject to further refinement and development, culminating in the paper by Bailey *et al.*[2] Further contributions to the study of transverse cracking have been given by a number of authors. For a reasonably complete account see Laws and Dvorak.[3]

The aim of the present paper is to give a succinct account of the progressive transverse cracking theory developed elsewhere by the present authors. The mechanical model described is, perhaps, the simplest one-dimensional model possible. Indeed, some straightforward stress analysis enables us to predict the loss of stiffness for a *given* crack density. This prediction compares well with experiment[3] and with the self-consistent model of Laws and Dvorak[4] and the lower bound given by Hashin.[5] However, from a practical point of view, the crucial problem is to predict the transverse crack density as a function of the applied (monotonic) load. This problem is solved herein by first analysing the fracture mechanics of transverse cracking and second, by giving an analysis of the statistics of progressive cracking. The result is a formula which permits us to predict transverse crack density as a function of applied load. Theory is shown to compare well with experiment.

2. PRELIMINARIES

In this paper we consider monotonic tensile loading of cross-ply composite laminates, see Fig. 1. It is convenient to use the subscripts t and *l* to refer to the transverse and longitudinal plies respectively. Thus the initial stresses in the laminate are respectively σ_t^R and σ_l^R.

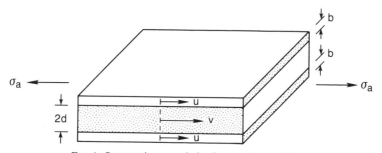

FIG. 1. Symmetric cross-ply laminate under axial load.

The one-dimensional theory which is outlined here assumes that the displacement of each layer is constant over the thickness of that layer. As usual, the displacements $u(x)$ and $v(x)$ are measured from the state in which there is no mechanical loading. The associated strains are

$$\varepsilon_t = \frac{dv}{dx}, \quad \varepsilon_l = \frac{du}{dx} \tag{1}$$

When the laminate is subject to mechanical loads, the total stress is the sum of the residual stress plus the stress due to loading:

$$\sigma_t = \sigma_t^R + \tau_t, \quad \tau_t = E_t \varepsilon_t \tag{2}$$

$$\sigma_l = \sigma_l^R + \tau_l, \quad \tau_l = E_l \varepsilon_l \tag{3}$$

where E denotes Young's modulus.

Turning now to the equilibrium of the laminate it is clear that

$$b\sigma_l^R + d\sigma_t^R = 0 \tag{4}$$

and that under applied stress σ_a

$$b\sigma_l + d\sigma_t = (b+d)\sigma_a \tag{5}$$

The essential feature of shear lag theory is that the shear stress which is responsible for the load transfer between the $0°$ and $90°$ plies is proportional to the relative displacement of the two layers. In other words

$$\tau = K(v - u) \tag{6}$$

where K is a constant.

Finally we recall from Ref. 3 that equilibrium of the $0°$ plies requires that

$$\tau = -b \frac{d\sigma_l}{dx} \tag{7}$$

whereas equilibrium of the $90°$ plies demands that

$$\tau = d \frac{d\sigma_t}{dx} \tag{8}$$

Within the framework of the theory developed here, Young's modulus for the uncracked laminate is given by

$$E_0 = \frac{bE_l + dE_t}{b+d} \tag{9}$$

Stress analysis of the cracked laminate is most easily achieved with the help of the differential equation for the transverse stress. This equation is obtained by differentiating eqn. (8) and making use of (1)–(6) together with (9) to give

$$\frac{d^2\sigma_t}{dx^2} - \frac{\xi^2}{d^2}\sigma_t = -\frac{\xi^2}{d^2}\left(\sigma_t^R + \frac{E_t}{E_0}\sigma_a\right) \tag{10}$$

Here, we have introduced the non-dimensional shear lag parameter

ξ through

$$\xi^2 = \frac{Kd(bE_l + dE_t)}{bE_l E_t} \qquad (11)$$

We now focus our attention on the ligament between adjacent transverse cracks, see Fig. 2. In this configuration, the differential equation (10) holds in the ligament AB and the correct boundary conditions for eqn. (10) are

$$\sigma_t = 0 \quad \text{when} \quad x = \pm h \qquad (12)$$

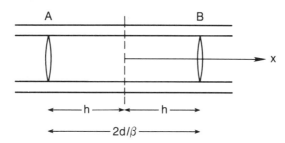

FIG. 2. Two adjacent transverse cracks in the 90° plies.

The solution for the transverse stress is

$$\sigma_t = \left(\sigma_t^R + \frac{E_t}{E_0} \sigma_a \right) \left(1 - \frac{\cosh \dfrac{\xi x}{d}}{\cosh \dfrac{\xi h}{d}} \right) \qquad (13)$$

As is shown by Laws and Dvorak,[3] it is not difficult to evaluate the displacements of the respective plies. For our present purposes it suffices to note that the average strain, ε_a, of the ligament AB, as measured at the surface of the laminate is

$$\varepsilon_a = \frac{\sigma_a}{E_0} \left\{ 1 + \frac{d^2 E_t}{\xi h b E_l} \tanh \frac{\xi h}{d} \right\} + \frac{d^2 \sigma_t^R}{\xi h b E_l} \tanh \frac{\xi h}{d} \qquad (14)$$

We see from (14) that the ligament has acquired a permanent strain, ε_p, given by

$$\varepsilon_p = \frac{d^2 \sigma_t^R}{\xi h b E_l} \tanh \frac{\xi h}{d} \qquad (15)$$

Clearly the permanent strain (15) is due to initial stress. However, it turns out that for all practical problems of interest, the value of ε_p is insignificant compared with ε_a. This simple theoretical argument supports the observation that any permanent strain, due to the relaxation of initial stress at crack surfaces, is negligible. Now, for a cracked laminate, in which the transverse crack density is β, the average distance between cracks is, from Fig. 2,

$$2h = 2d/\beta$$

Hence from (14) we obtain the formula for the Young's modulus $E(\beta)$, of the cracked laminate:

$$E(\beta) = E_0 \left\{ 1 + \frac{\beta}{\xi} \frac{dE_t}{bE_l} \tanh \frac{\xi}{\beta} \right\}^{-1} \tag{16}$$

It is clear from (16) that in order to predict the loss of stiffness of a cross-ply laminate we need to know the value of the shear lag parameter ξ. This has been argued in depth by Laws and Dvorak.[3] Briefly, the suggestion is that ξ be *determined* by measurement of the first ply failure stress — as is discussed shortly. It turns out that the loss of stiffness is not particularly sensitive to the value of ξ. It is also appropriate to mention that the self-consistent model of Laws and Dvorak,[4] the lower bound obtained by Hashin,[5] as well as the shear lag model[3] give excellent agreement with experiment. For details we refer the interested reader to Ref. 3.

In this contribution we prefer to focus on the more difficult problem of predicting the transverse crack density (β) as a function of the applied load (σ_a). Once this relationship is available it is clear that we can also predict the appropriate stress–strain relation, etc.

3. PROGRESSIVE CRACKING

The detailed analysis of progressive transverse cracking is given by Laws and Dvorak[3] so we are content here to emphasize the main results. Consider the uncracked ligament of Fig. 2. When the applied load reaches a critical value this ligament will itself crack at some location C, as in Fig. 3. Since the transverse cracking process is not deterministic there is no reason to suppose that C lies at the mid-point of AB. It therefore follows that we need to compute the energy release rate for a crack at an arbitrary location C propagating across the ply. The required calculation is not easy. Nevertheless we can read off the required result from the paper of Laws and

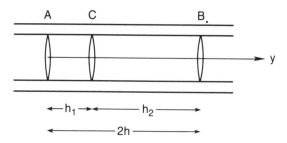

FIG. 3. The ligament AB with an additional crack at C.

Dvorak:[3]

$$G = \frac{d(b+d)E_0}{\xi b E_l E_t} \left(\sigma_t^R + \frac{E_t}{E_0} \sigma_a \right)^2 \left\{ \tanh \frac{\xi h_1}{2d} + \tanh \frac{\xi h_2}{2d} - \tanh \frac{\xi h}{d} \right\} \quad (17)$$

To get the required energy release rate at first ply failure we consider the limit in which h, h_1 and $h_2 \to \infty$, namely

$$G^{fpf} = \frac{d(b+d)E_0}{\xi b E_l E_t} \left(\sigma_t^R + \frac{E_t}{E_0} \sigma_a^{fpf} \right)^2$$

Clearly we get first ply failure when

$$G^{fpf} = G_c$$

Thus it follows that

$$\xi = \frac{d(b+d)E_0}{G_c b E_l E_t} \left(\sigma_t^R + \frac{E_t}{E_0} \sigma_a^{fpf} \right)^2 \quad (18)$$

Now standard data for cross-ply laminates is given by Wang[6] for two different graphite–epoxy systems. In short, all quantities in eqn. (18), except ξ, are either given or easily inferred. Thus we regard eqn. (18) as the rule which determines the shear lag parameter ξ. Values for specific systems are given later.

We now turn to the problem of determining crack density under increasing load. Here it is necessary to make some assumptions about the statistics of progressive cracking. In this connection we refer to some arguments of Laws and Dvorak[3] who appealed to elementary fracture mechanics to show that an appropriate choice for the probability density function, p, for the site of the next crack in the ligament of Fig. 3, is that p be proportional to the stress in the transverse ply. Of course, normalization gives the factor of proportionality. Thus from eqn. (13) with x replaced by

$(y-h)$ we find

$$p(y)= \frac{1}{2h} \left\{ 1 - \frac{\cosh \dfrac{\xi(y-h)}{d}}{\cosh \dfrac{\xi h}{d}} \right\} \left\{ 1 - \frac{\tanh \dfrac{\xi h}{d}}{\dfrac{\xi h}{d}} \right\}^{-1} \tag{19}$$

It now follows that in a laminate which already contains cracks of density β, the expected value of the applied stress to cause additional cracking is

$$E(\sigma_a(\beta)) = \int_0^{2h} p(y)\, \sigma_a(y)\, \mathrm{d}y \tag{20}$$

The integral in eqn. (20) must be evaluated numerically and demands considerable care.

We now compare the theoretical predictions with some experimental results of Crossman *et al.*[7] for two graphite–epoxy systems. Perhaps the most convenient sources for the data are the survey article by Wang[8] together with the report.[6] Use of these data enables us to calculate the appropriate value of ξ directly from eqn. (18). The results are, *cf.* Ref. 3,

$$(0_2, \ 90)_s \qquad \xi = 0.93$$
$$(0_2, \ 90_2)_s \qquad \xi = 1.38$$
$$(0_2, \ 90_3)_s \qquad \xi = 2.24$$

Comparisons of the theoretical predictions with experimental results are shown in Fig. 4. For clarity we have not included the predictions of the

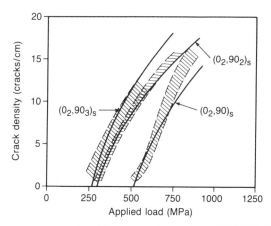

FIG. 4. Theory versus experiment for progressive cracking of AS-3501-06 laminates.

Wang–Crossman theory in Fig. 4. However, we note that the theory of these authors also gives excellent agreement.

Finally, we turn to some experimental data obtained by Wang[8] for some T300/934 laminates. For these laminates we infer the values of ξ to be as follows:[3]

$$(0, \quad 90_2, \quad 0) \qquad \xi = 1 \cdot 08$$
$$(0, \quad 90_3, \quad 0) \qquad \xi = 1 \cdot 70$$
$$(0, \quad 90_4, \quad 0) \qquad \xi = 1 \cdot 79$$

Theory is compared with experiment in Fig. 5. Agreement is good.

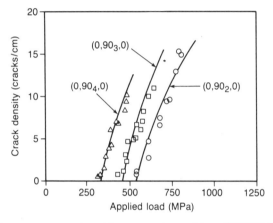

FIG. 5. Theory versus experiment for progressive cracking of T300/934 laminates.

ACKNOWLEDGEMENTS

This work was supported by a grant from the Air Force Office of Scientific Research. Mr W. Li helped in the interpretation of the experimental data and computing.

REFERENCES

1. GARRETT, K. W. and BAILEY, J. E., Multiple transverse fracture in 90° cross-ply laminates of glass fibre-reinforced polyester, J. Mater. Sci., 12 (1977), 157.
2. BAILEY, J. E., CURTIS, P. T. and PARVIZI, A., On the transverse cracking and longitudinal splitting behaviour of glass and carbon fibre reinforced epoxy cross

ply laminates and the effect of Poisson and thermally generated strain, *Proc. R. Soc. Lond. A.*, **366** (1979), 599.

3. LAWS, N. and DVORAK, G. J., Stiffness reduction and progressive matrix cracking in composite laminates under monotonic loading, *J Composite Mater.*

4. LAWS, N. and DVORAK, G. J., The loss of stiffness of cracked laminates, *Proc. IUTAM Eshelby Memorial Symposium*, Cambridge University Press, 1985, p. 119.

5. HASHIN, Z., Analysis of cracked laminates: a variational approach, *Mech. Mater.*, **4** (1985), 121.

6. WANG, A. S. D., Fracture analysis of matrix cracking in laminated composites, Report No. NADC-85118-60, Drexel University, 1985.

7. CROSSMAN, F. W., WARREN, W. J., WANG, A. S. D. and LAW, G. E., Initiation and growth of transverse cracks and edge delamination in composite laminates, part 2: experimental correlation, *J. Comp. Mater. Suppl.*, **14** (1980), 88.

8. WANG, A. S. D., Fracture mechanics of sublaminate cracks in composite materials, *Compos. Technol. Rev.*, **6** (1984), 45.

7

Singularities in Composite Materials Applications

R. S. Barsoum

*US Army Materials Technology Laboratory,
Watertown, Massachusetts, USA*

ABSTRACT

Predicting failure modes and mechanisms of failure in composites hinges to a great extent on our knowledge of the stress fields associated with discontinuities in composites. In composites, singularities arise at discontinuities at the micro and macro scale of their behavior, and whence affect the failure mode and damage at all levels. Such discontinuities include interfaces, boundaries of laminated composites, cracks between layers, design discontinuities such as joints, cut outs, cracks, repairs, etc. The singularities in many of these cases are not known, or the known solutions are too complex, and no general method is at the disposal of the designer or experimentalist for analysing the specific case he is dealing with.

Recently, the author developed a general Finite Element based iterative method for the solution of eigenvalue problems common in the fracture of composite materials. The advantage of the method is that it relies on the use of general purpose finite element packages for performing the iterative analysis. The data are processed to evaluate both power singularities and oscillating singularities, occurring at interface cracks of composites. In this presentation, we will illustrate several problems of interface cracks, and delamination, and show how the iterative finite element method is used in evaluating the singular field. The implications of the singularities to failure mechanisms of fiber/ matrix debonding, edge delamination and the behavior of laminated composites will be discussed.

1. SINGULARITY FIELDS IN COMPOSITE MATERIALS

The asymptotic field in many singularity problems associated with the fracture of composite materials can be written as[1]

$$\sigma_{ij} = \frac{K_1}{r^\delta} f^1_{ij}(\theta, C_k) + \frac{K_2}{r^\delta} f^2_{ij}(\theta, C_k) \tag{1a}$$

and

$$u_i = \frac{K_1}{E^*} r^{1-\delta} F^1_i(\theta, C_k) + \frac{K_2}{E^*} r^{1-\delta} F^2_i(\theta, C_k) \tag{1b}$$

where δ is the power of the singularity, which could be real or complex depending on the nature of the discontinuity and material properties. The functions f^n_{ij} and F^n_i are also dependent on the materials properties C_k and are both bounded functions. The restrictions on the power of singularity are that it produces finite displacements at the origin and that the resulting strain energy is positive definite.

1.1. The Finite Element Iterative Method

The aim of the finite element iterative approach discussed here is the evaluation of singularity δ and the functions $f_{ij}(\theta)$ and $F_i(\theta)$ in eqn. (1). It will also be shown that the approach is a global method and whence the stress intensities can be calculated. This technique will be referred to as the Finite Element Iterative Method (FEIM).[2-5] Since the method uses a displacement formulation with elements that satisfy all the requirements of rigid body motion, constant strain, positive definite strain energy, and finite displacement for linear elastic problems, the conditions on the asymptotic eigenvalue solution are satisfied *a priori*.

1.1.1. Procedure

FEIM uses existing general purpose Finite Element Programs as follows:

1. Construct a circular mesh around the crack or singularity, Fig. 1.
2. The radii of the rings should follow an (r^2) distribution. For crack problems the quarter-point elements are used to represent the square root singularity.
3. Impose arbitrary boundary displacements on the outer boundary R_b —an approximate solution will accelerate the convergence.
4. Choose radius R_s near the crack tip or singularity, where the displacements $\{u_{R_s}\}$ are evaluated.

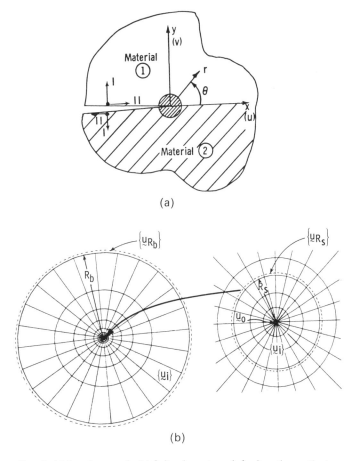

(a)

(b)

FIG. 1. (a) Interface crack; (b) finite element mesh for iterative method.

5. Perform a finite element analysis and extract the displacements $\{\mathbf{u}_{R_s}\}$. Scale the results at R_s and apply these as boundary displacements $\{\mathbf{u}_{R_b}\}$ at the outer boundary R_b.
6. Repeat step 5 until convergence is attained.
7. To improve the results, subtract the crack tip displacements $\mathbf{u}_o$ from the displacements at the rind R_s before scaling. The displacements $\mathbf{u}_o$ represent a rigid body motion and therefore do not affect the solution.

The above procedure was originally applied to the case of a crack in a general anisotropic material[2] and was found to converge to the asymptotic field exact solution.[6] The singularity in this case is $\sqrt{r}$, and the only unknowns are the distribution functions $f_{ij}(\theta)$ and $F_i(\theta)$.

Convergence is judged when the displacements at any ring eventually approach $r^{1-\delta} F_i(\theta)$. Hence, δ and $F_i(\theta)$ can be extracted from the analysis. In the cases where the singularity δ is a real number, we find that the displacements $\{u_{R_s}\}$ do not change from one iteration to the next, and the scaling parameter reaches a constant value. The asymptotic field in this case is called self-similar. In the case where the singularity δ is a complex member, the displacements $\{u_{R_s}\}$ have a phase shift from one iteration to the next and the scaling parameter could oscillate wildly from one iteration to the next. However, that is the nature of the problem and the evaluation of this field, which is called a non-self-similar field, will be discussed later.

2. IDENTIFICATION OF THE FEIM WITH THE POWER METHOD FOR EIGENVALUE PROBLEMS

The Finite Element Iterative Method was identified with the Power Sweep method for the determination of eigenvalues.[4] We will show here that the nature of the iterative procedure for evaluating the eigenvalues is actually a minimization of the Rayleigh quotient.

The equations of equilibrium solved in each step are given by

$$\begin{bmatrix} K_i & | & K_{ib} \\ --- & | & --- \\ K_{bi} & | & K_{bb} \end{bmatrix} \begin{Bmatrix} u_i \\ --- \\ u_{R_b} \end{Bmatrix} = \{0\} \tag{2}$$

where $\{u_i\}$ represents all displacements other than those on the outer boundary R_b. During the iterative process, we solve

$$\{u_i^n\} = -[K_i]^{-1}[K_{ib}]\{u_{R_b}^n\} \tag{3}$$

where

$$\vdots$$

$$\{u_{R_b}^{n+1}\} = \alpha(\{u_{R_s}^n\} - \{u_0^n\}) \tag{4}$$

$$\vdots$$

By further partitioning of eqn. (2), $\{u_i\}^T = \lfloor u_o, u_{R_s}, u_{R_i} \rfloor$ it is possible[4] to write eqn. (3) as

$$[A]\{u_{R_s}^n\} = -[B]\{u_{R_b}^n\} \tag{5}$$

where $[A]$ and $[B]$ are square matrices. Therefore, eqns (4) and (5) form a statement of the generalized eigenvalue problem.[7] Inverting the matrix $[A]$, we can write eqns (4) and (5) as

$$\{u_{R_s}^{n+1}\} = \lambda [T]\{u_{R_s}^n\} \tag{6}$$

where the eigenvalue λ can be identified with the radial function part of eqn. (1b), i.e. $r^{(1-\delta)}$. Therefore,

$$\lambda = (R_b/R_s)^{(1-\delta)} \tag{7}$$

The displacements $\{u_{R_s}\}$ or eigenfunctions of eqn. (6), are identified with the angular distribution of the asymptotic field, thus

$$\{u_{R_s}\} = F_i^n(\theta, C_k) \tag{8}$$

The matrix $[T]$ is referred to as the transfer matrix.[8] It is possible to show that when the matrix $[T]$ is symmetric, the singularity is real and when it is non-symmetric, the singularity δ is complex.

Although we have identified the eigenvalue with a power type singularity in eqn. (7), there is no reason not to assume that

$$\lambda = L(r) \tag{9}$$

where the asymptotic field of eqn. (1b) can be written in the separable form,

$$u_i(r, \theta) = L(r)F_i^1(\theta) + L(r)F_i^2(\theta) \tag{10}$$

The function $L(r)$ could be logarithmic or power singularity or a combination. As long as the field is separable, we can find out the $L(r)$ function by fitting it to the displacements along any radial line in Fig. 1.

2.1. Complex Eigenvalues

Complex eigenvalue problems arise when the transfer matrix $[T]$ is non-symmetric. After a large number of iterations (n), one can write eqn. (6) as

$$[T]^n\{u_{R_b}^o\} = \lambda_1^n \alpha_1 x_1 + \bar{\lambda}_1^n \bar{\alpha}_1 \bar{x}_1 + \sum_3^N \alpha_i x_i \lambda_i^n \tag{11}$$

where $x_j, j = 1, \ldots, N$ are conjugate complex eigenfunctions of the problem. They represent a complete set N for the matrix $[T]$ (order N). $x_1, \bar{x}_1$ are the corresponding complex conjugate eigenvectors of the first dominant eigenvalue λ_1, and its conjugate $\bar{\lambda}_1$. α_i are conjugate constants.

Using the Rayleigh quotient approach, it is possible to show that the last summation in eqn. (10), will get small in comparison with the first two terms as the number of iterations gets larger. Therefore, after a large number of iterations, the method will converge to the first dominant mode, which is represented by the first two terms of eqn. (11).

At convergence we can also write the following equation,

$$\beta_1\{\mathbf{u}_{R_s}^n\} + \beta_2\{\mathbf{u}_{R_s}^{n+1}\} + \beta_3\{\mathbf{u}_{R_s}^{n+2}\} = 0 \tag{12}$$

where β_i can be determined by least squares method for the over determined system of equations. The author determined β_i by satisfying eqn. (12) in the sense of the norms, i.e. dot products are used. These equations were solved to calculate β_1, β_2 and β_3, each equation was obtained by multiplying eqn. (12) by one of the iteration vectors.

Using eqns (11) and (12), we get the characteristic equation for λ, given by:

$$\beta_1 + \beta_2\lambda_1 + \beta_3\lambda_1^2 = 0 \tag{13}$$

from which λ_1 can be obtained and the singularity can be calculated from eqn. (7) or (9). For a power singularity

$$\lambda_1 = \xi_1 + i\eta_1 = (R_b/R_s)^{a+i\varepsilon} \tag{14}$$

or

$$\xi_1 + i\eta_1 = (R_b/R_s)^a[\cos(\varepsilon \ln(R_b/R_s)) + i\sin(\varepsilon \ln(R_b/R_s))] \tag{15}$$

from which the real and imaginary parts, a and ε, can be evaluated.

The dominant eigenvector $\mathbf{x}_1$ can be calculated from two consecutive iterations, (n) and $(n+1)$. Therefore,

$$\mathbf{x}_1 = \eta_1\{\mathbf{u}_{R_s}^n\} + i(\xi_1\{\mathbf{u}_{R_s}^n\} - k_{n+1}\{\mathbf{u}_{R_s}^{n+1}\}) \tag{16}$$

where k_{n+1} is the scaling factor during that iteration.

2.2. Global Interpretation of FEIM

The FEIM can also be interpreted as a substructuring method. Each iteration represents a substructure at the radius R_s. This means that the iterations are continuously zeroing-in on the singular point (e.g. crack tip) in every successive analysis. It can be shown that the stiffness from iteration (n) to iteration $(n+1)$ should be scaled in the following manner,

$$[K]^{n+1} = (R_s/R_b)^2[K]^n \tag{17}$$

Therefore, if the initial displacements $\{\mathbf{u}_{R_b}^o\}$ were obtained from those of a model representing a structure or test specimen, the FEIM, in addition to evaluating the singularity field, will also give the stress intensity factors. Of course, proper multiplications of the results using eqn. (17) will be needed.

For the case of complex eigenvalues, the stress intensities are calculated from eqns (16) and (1), therefore

$$\mathbf{x} = KF_\alpha(\theta) \quad \text{and} \quad \alpha = x, y \tag{18a}$$

where the stress intensity K, in this case is complex. It can be written as;

$$K = (K_1 + iK_2) \quad \text{and} \quad F_\alpha = \mathbf{f} + i\mathbf{g} \tag{18b}$$

If we also write $\mathbf{x}$ in its real and imaginary parts, we get,

$$\mathbf{x} = \mathbf{z} + i\mathbf{w} = (K_1 + iK_2)(\mathbf{f} + i\mathbf{g}) \tag{19}$$

Using dot products we get

$$K_1 = (\mathbf{z} \cdot \mathbf{f} + \mathbf{w} \cdot \mathbf{g})/(\mathbf{f} \cdot \mathbf{f} + \mathbf{g} \cdot \mathbf{g}) \tag{20a}$$

$$K_2 = -(\mathbf{z} \cdot \mathbf{g} + \mathbf{w} \cdot \mathbf{f})/(\mathbf{f} \cdot \mathbf{f} + \mathbf{g} \cdot \mathbf{g}) \tag{20b}$$

For real eigenvalues the above equations reduce to

$$\begin{aligned} K_1 &= (\mathbf{z} \cdot \mathbf{f})/(\mathbf{f} \cdot \mathbf{f}) \\ K_2 &= 0 \end{aligned} \tag{21}$$

where $\mathbf{f}$ is the analytical expression for the functional variation in θ for the mode of fracture in interest (I, II, or III). Equations (20) and (21) represent a powerful method for evaluating stress intensities for complex and real asymptotic fields. It should be noted that its accuracy is like integral methods.

2.3. Verification of FEIM

The FEIM was verified for cracks in anisotropic materials.[2] Tests on singularities at interfaces included cracks perpendicular to interfaces,[4] free surface meeting an interface,[4] and crack along an interface of dissimilar media.[5]

The results in all these cases, where analytical solutions were available, were extremely accurate.

The most severe test[5] was that of the crack along the interface of dissimilar media.[9-12] In that case, the real and imaginary parts of eqn. (14) for the power singularity were given by

$$a + i\varepsilon = 0.500\ 004\ 3 - i(0.075\ 786\ 03) \tag{22}$$

The analytical solution[12] for dissimilar materials with properties

$$G_1/G_2 = 1/10, \ v_1 = v_2 = 0.3$$

is given by

$$a + i\varepsilon = 0 \cdot 5 - i(0 \cdot 075\ 811\ 78) \tag{23}$$

where ε is calculated from the equation,

$$\varepsilon = \frac{1}{2\pi} \ln \left[\frac{(3 - 4v_1)/G_1 + 1/G_2}{(3 - 4v_2)/G_2 + 1/G_1} \right] \tag{24}$$

2.4. Applications

The Finite Element Iterative Method (FEIM) as outlined here exhibits all of the generalities of the finite element method. Thus problems involving two- and three-dimensional singularities as well as various material anisotropy are easily accommodated. Problems of singularity at interfaces of isotropic as well as anisotropic materials can be handled easily with the method using the complex eigenvalue method for the determination of the asymptotic field. It was shown that as long as the field can be written in a separable form for the r and θ functions, the method will work.

Solutions of several problems in composite materials are underway. These include singularities at delaminations,[13] free edge singularities[14] at cut-outs and lap joints, and intersection of a crack between dissimilar media with the free surface. The latter problem was solved for homogeneous isotropic media.[15,16] However, some of these results are not in agreement with the rest.

3. CONCLUSIONS

It was shown that the finite element iterative method is a statement of the eigenvalue problem of the asymptotic singular field. The FEIM uses existing general purpose Finite Element Programs with little or no modification, and therefore represents a powerful tool for the practical engineer. The method has all the generalities of the finite element method for 2-D, 3-D, material anisotropy, and orientation.

We have also shown that the method is also a global method, and whence the stress intensities for real as well as complex fields can be calculated.

REFERENCES

1. ERDOGAN, F., Stress intensity factors, *ASME Journal of Applied Mechanics*, **49** (1982), 561–69.

2. BARSOUM, R. S., Cracks in anisotropic materials — an iterative solution of the eigenvalue problem, *International Journal of Fracture*, **32** (1986), 59–67.

3. BARSOUM, R. S., Finite Element Solution of the Eigenvalue Problems Associated with the Fracture of Composites, *Invited Lecture*, Fourth International Conference on Numerical Methods in Fracture Mechanics, San Antonio, TX, 23–27 March 1987, Swansea, UK, Pineridge Press Ltd, 1987.

4. BARSOUM, R. S., Theoretical basis of the finite element iterative method for the eigenvalue problem in stationary cracks, *International Journal of Numerical Methods in Engineering*, **25** (1988).

5. BARSOUM, R. S., Application of the finite element iterative method to the eigenvalue problem of a crack between dissimilar media, *International Journal of Numerical Methods in Engineering*, **25** (1988).

6. SIH, G. C. and CHEN, E. P., *Mechanics of Fracture – Cracks in Composite Materials*, The Hague, Martinus Nijhoff, 1981.

7. WILKINSON, J. H., *The Algebraic Eigenvalue Problem*, Oxford, Clarendon Press, 1965.

8. LIVESLY, R. K., *Matrix Methods of Structural Analysis*, Oxford, Pergamon Press, 1964.

9. WILLIAMS, M. L., The stress around a fault or crack in dissimilar media, *Bulletin of the Seismological Society of America*, **49** (1959), 199–204.

10. ERDOGAN, F., Stress distribution in a non-homogeneous elastic plane with cracks, *International Journal of Applied Mechanics, Trans. ASME, Series E*, **85** (1963), 232–7.

11. SIH, G. C. and RICE, J. R., The bending of plates of dissimilar materials with cracks, *Journal of Applied Mechanics, Trans. ASME*, **31** (1964), 477–82.

12. RICE, J. R. and SIH, G. C., Plane problems of cracks in dissimilar media, *Journal of Applied Mechanics, Trans. ASME*, **32** (1965), 418–23.

13. WANG, S. S. and CHOI, I., The interface crack behavior in dissimilar anisotropic composites under mixed-mode loading, *Journal of Applied Mechanics, ASME*, **50** (1983), 179–83.

14. ZWIERS, R. I., TING, T. C. T. and SPILKER, R. L., On the logarithmic singularity of free edge stress in laminated composites under uniform extension, *Journal of Applied Mechanics, ASME*, **49** (1982), 561–9.

15. BAZANT, P. Z. and ESTENSSORO, L. F., Surface singularity and crack propagation, *International Journal of Solids and Structures*, **15**, (1979), 405–26.

16. BENTHEM, J. P., State of stress at the vertex of a quarter-infinite crack in a half-space, *International Journal of Solids and Structures*, **13**, (1977), 479–92.

8

Transferability of Small Specimen Data to Full-Size Structural Components

A. CARPINTERI

Department of Structural Engineering, Politecnico di Torino, Turin, Italy

and

P. BOCCA

Istituto Universitario di Architettura di Venezia, Venice, Italy

ABSTRACT

The transition from slow crack growth (ductile behaviour) to rapid crack propagation (brittle behaviour) by varying the structural size scale and keeping the structural shape constant, is explained in terms of dimensional analysis and described with the application of the Strain Energy Density Theory. On the other hand, in spite of the apparent variability of the constitutive behaviour as a function of size of material element, the damage mechanics is proposed as an invariant process at the microscale, which can be revealed experimentally by temperature fluctuations.

1. INTRODUCTION

In the last few years a great effort has been made by researchers and institutions to explain two peculiar and recurrent phenomena in material strength:

(1) size effect;
(2) stable crack growth.

As regards the former, if the structural size scale varies, the structural shape being constant, the mechanical behaviour of the structure decidedly

changes from the very brittle to the very ductile (Fig. 1). This effect can be
explained only by the application of physical similitude and scale modelling
concepts.[1-5] As regards the latter phenomenon, it represents only a local
instability at the crack tip and its nature is totally different from that of
unstable crack propagation.[6]

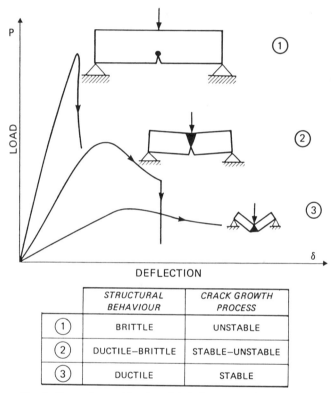

	STRUCTURAL BEHAVIOUR	CRACK GROWTH PROCESS
①	BRITTLE	UNSTABLE
②	DUCTILE–BRITTLE	STABLE–UNSTABLE
③	DUCTILE	STABLE

FIG. 1. Size-scale transition from brittle fracture to plastic collapse.

Stable crack growth may occur both under monotonic or repeated
loading and may precede or follow the unstable crack propagation. The
fundamental laws governing the transition from slow to rapid crack
propagation, and vice versa, should be very general and applicable to very
simple as well as to very complex structures, so that it will be easy and
consistent to extrapolate the results obtained from small specimens to the
project of large structures.

The size-scale transition from plastic collapse to brittle fracture will be
examined on the basis of dimensional analysis and described by the

application of the Strain Energy Density Theory.[7,8]. Then, the size effects on material strength, toughness and ductility will be discussed through the scale-invariance of the non-dimensional crack growth resistance curves. Eventually, the cooling-heating effect in a heterogeneous material subjected to repeated loading will be introduced and explained as a transition from order to disorder (or damage) according to the thermodynamics of irreversible processes.

2. DIMENSIONAL TRANSITION FROM PLASTIC COLLAPSE TO BRITTLE FRACTURE

Two fundamental questions arise dramatically.

(1) Are the data coming from small-scale specimens related to the collapse conditions in large-scale structures?

(2) If the ductile fracture in small-scale specimens is not completely obscured by the plastic flow collapse at the ligament, how is it possible to put the former in connection with the brittle fracture in large-scale structures?

The competition between collapses of a different nature can be emphasized with the application of dimensional analysis and considering the maximum loads derived from LEFM and limit analysis respectively. The transition from ductile to brittle behaviour is governed by a non-dimensional brittleness number which is a function of material properties and structure size scale. A true separation collapse occurs only with relatively low fracture toughness and/or large structure size.

Due to the different physical dimensions of yield strength, σ_y, and fracture toughness, K_{IC}, scale effects are always present in the usual fracture testing of common engineering materials. This means that, for the usual size scale of the laboratory specimen, the plastic collapse at the ligament tends to anticipate and obscure the brittle crack propagation. Such a competition between different types of collapses can easily be shown by considering the expression for the stress-intensity factor in a centre cracked slab (Fig. 2):

$$K_1 = \sigma \sqrt{\pi a_o}\, f\left(\frac{a_o}{b}\right) \tag{1}$$

where the shape function f is:

$$f\left(\frac{a_o}{b}\right) = \left(\sec\frac{\pi a_o}{2b}\right)^{1/2}$$

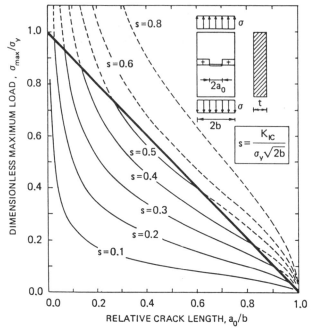

FIG. 2. Interaction between brittle crack propagation and plastic collapse, by varying the brittleness number $s = K_{IC}/\sigma_y\sqrt{2b}$.

At the crack propagation condition eqn. (1) becomes:

$$K_{IC} = \sigma_{max} \sqrt{\pi a_o}\, f\left(\frac{a_o}{b}\right) \tag{2}$$

If both members of eqn. (2) are divided by $\sigma_y\sqrt{2b}$, we obtain:

$$\frac{K_{IC}}{\sigma_y\sqrt{2b}} = s = \frac{\sigma_{max}}{\sigma_y} \sqrt{\frac{\pi a_o}{2b}}\, f\left(\frac{a_o}{b}\right) \tag{3}$$

where s is a dimensionless number able to describe the brittleness of the system and where both material properties and specimen size appear. Rearranging eqn. (3) gives:

$$\frac{\sigma_{max}}{\sigma_y} = s \left(\frac{\cos\dfrac{\pi a_o}{2b}}{\dfrac{\pi a_o}{2b}}\right)^{1/2} \tag{4}$$

On the other hand, it is possible to consider even the non-dimensional load producing plastic collapse at the ligament $(b - a_o)$:

$$\frac{\sigma_{max}}{\sigma_y} = 1 - \frac{a_o}{b} \tag{5}$$

Equations (4) and (5) are plotted in Fig. 2 as functions of the crack depth a_o/b. While the former provides a family of curves by varying the brittleness number s, the latter is represented by a unique curve (thick line). It is easy to realize that plastic collapse at the ligament precedes crack propagation for each initial crack depth when the brittleness number s is higher than the critical value $s_o = 0.54$. For lower s numbers, plastic collapse anticipates crack propagation only for initial crack depths external to a certain interval. This means that a real separation phenomenon occurs only with a relatively low fracture toughness, high yield strength and/or large structure size. Not the single values of K_{IC}, σ_y and b, but only their function s — see eqn. (3) — determines the nature of the collapse mechanism.

3. STRAIN ENERGY DENSITY THEORY

3.1. Mechanical Damage and Strain-Softening Behaviour

Damage of the material at the crack tip and crack growth increments will be computed on the basis of a uniaxial bilinear elastic-softening stress-strain relation (Fig. 3(a)). If the loading is relaxed when the representative point is in A, the unloading is assumed to occur along the line AO, so that the new bilinear constitutive relation is the line OAF. No permanent deformation is allowed by such a model, but only the degradation of the elastic modulus. The present model simulates the mechanical damage by decreasing elastic modulus, E, and strain energy density which can be absorbed by a material element. In fact, while for a non-damaged material element the critical value of the strain energy density, $(dW/dV)_c$, is equal to the area OUF (Fig. 3(a)) for a damaged material element with representative point in A, the decreased critical value, $(dW/dV)_c^*$, is equal to the area OAF. In addition, as is shown in Fig. 3(a), the area OUA represents the dissipated strain energy density, $(dW/dV)_d$, OAB the recoverable strain energy density, $(dW/dV)_r$, and BAF the additional strain energy density, $(dW/dV)_a$.

The described model will be extended to the three-dimensional stress conditions, using the current value of the absorbed strain energy density, (dW/dV), as a measure of damage. In other words, the effective elastic modulus, E^*, and the decreased critical value of strain energy density,

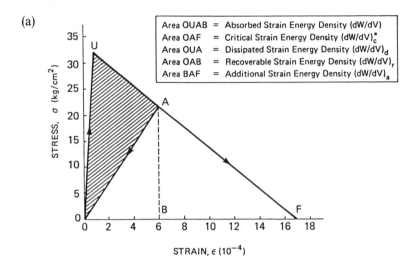

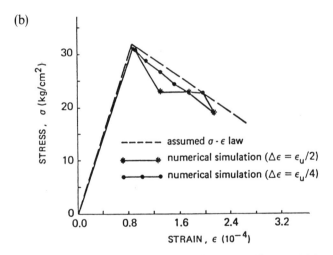

FIG. 3. Strain-softening constitutive law of a concrete-like material in tension: (a) assumed bilinear relation; (b) numerical damage simulation.

$(dW/dV)_c^*$, will be considered as functions of the absorbed strain energy density, (dW/dV), being:

$$\left(\frac{dW}{dV}\right) = \left(\frac{dW}{dV}\right)_d + \left(\frac{dW}{dV}\right)_r$$

Such functions in the uniaxial case are:

$$\text{OUAB-area} \rightarrow \text{AO-slope, i.e.} \quad \left(\frac{dW}{dV}\right) \rightarrow E^* \tag{6a}$$

$$\text{OUAB-area} \rightarrow \text{OAF-area, i.e.} \quad \left(\frac{dW}{dV}\right) \rightarrow \left(\frac{dW}{dV}\right)_c^* \tag{6b}$$

Stress and strain in the softening condition A (Fig. 3(a)), can be expressed in terms of stress and strain in the ultimate and fracture conditions:

$$\sigma = E^* \varepsilon = \frac{\sigma_u \varepsilon_f}{(\varepsilon_f - \varepsilon_u) + \dfrac{\sigma_u}{E^*}} \tag{7}$$

Through eqns. (6) and (7) it is simple to express the absorbed strain energy density, (dW/dV), and the decreased critical value, $(dW/dV)_c^*$, as functions of the constitutive parameters, σ_u, ε_u, ε_f, and of the effective Young's modulus, E^*:

$$\left(\frac{dW}{dV}\right) = \frac{1}{2}(\sigma\varepsilon + \sigma_u\varepsilon - \sigma\varepsilon_u) \tag{8a}$$

$$\left(\frac{dW}{dV}\right)_r = \frac{1}{2}\sigma\varepsilon \tag{8b}$$

$$\left(\frac{dW}{dV}\right)_d = \left(\frac{dW}{dV}\right) - \left(\frac{dW}{dV}\right)_r = \frac{1}{2}(\sigma_u\varepsilon - \sigma\varepsilon_u) \tag{8c}$$

$$\left(\frac{dW}{dV}\right)_c^* = \left(\frac{dW}{dV}\right)_c - \left(\frac{dW}{dV}\right)_d = \frac{1}{2}(\sigma_u\varepsilon_f - \sigma_u\varepsilon + \sigma\varepsilon_u) \tag{8d}$$

The relations (8) will be discretized using 25 different values of the elastic modulus:

$$E^*(n) = \frac{(26-n)}{25}E, \quad \text{for} \quad n = 1, 2, \ldots, 25 \tag{9}$$

In Fig. 3(b), the results of different tensile test numerical simulations are reported. Two strain-controlled loading processes are carried out and the strain increment in one case is twice as in the other one. The discretized stress–strain relations are displayed and the comparison with the assumed σ–ε constitutive law is also shown. It can be proved that, when the effective elastic modulus, E^*, varies continuously and the strain increment $\Delta\varepsilon$ tends to zero, the assumed bilinear $\sigma - \varepsilon$ variation (dashed line of Fig. 3(b)) is exactly reproduced by the numerical damage simulation.

3.2. Slow Crack Growth versus Unstable Crack Propagation

In order to evaluate the crack growth increment at each loading step, the Strain Energy Density Theory will be applied, as proposed by Sih.[7,8] It is based on the following fundamental assumptions.

(1) The stress field in the vicinity of the crack tip cannot be described in analytical terms because of the relative heterogeneity of the material. A minimum distance, r_0, does exist below which it is a nonsense to study the mechanical behaviour of the material from a 'continuum mechanics' point of view and to consider macroscopic crack growth increments.

(2) Out of such a core region of radius r_0, the strain energy density field can always be described by means of the following general relationship:

$$\left(\frac{dW}{dV}\right) = \frac{S}{r} \tag{10}$$

where the strain energy density factor, S, is generally a function of the three space coordinates.

(3) According to Beltrami's criterion, all the material elements in front of the crack tip, where the strain energy density is higher than the critical value, $(dW/dV)_c^*$, fail.

(4) When the following condition holds:

$$\Delta a = r_0^* = S_0/(dW/dV)_c^* \tag{11}$$

the crack may be considered as arrested, at least from a macroscopical point of view. On the other hand, when the crack growth increment is:

$$\Delta a = r_c^* = S_c/(dW/dV)_c^* \tag{12}$$

the unstable crack propagation takes place. S_c is a material constant and represents the strength of the material against rapid and uncontrollable crack propagation. S_c is connected with the critical value of the stress-intensity factor, K_{IC}, through the following equation (plane strain condition).[7]

$$S_c = \frac{(1+v)(1-2v)}{2\pi E} K_{IC}^2 \tag{13}$$

3.3. Centre Cracked Slab in Tension

A centre cracked slab in tension (Fig. 4(a)) is analysed by using the Axisymmetric/Planar Elastic Structures (APES) finite element program.[9] It is a computer program which incorporates 12-noded quadrilateral isoparametric elements allowing for cubic displacement fields and quadratic stress and strain fields within each element. The r^{-1} strain energy density singularity in the vicinity of the crack tip is embedded in the solution through the use of $\frac{1}{9}-\frac{4}{9}$ nodal spacing on the element sides adjacent to the crack tip. The idealization of Fig. 4(b) utilizes 309 nodes and 52 elements and is considered in a condition of plane strain.

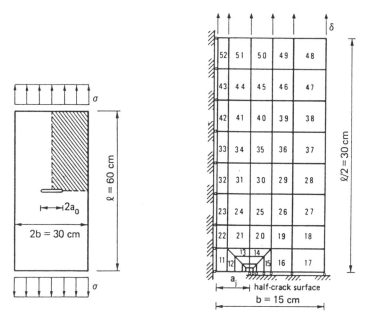

FIG. 4. Centre cracked slab in tension.

The strain energy density criterion is applied to the tension test specimen, considering a strain-controlled loading process. The stress–strain responses for three different initial crack lengths are displayed in Fig. 5. The load carrying capacity decreases by increasing the initial crack length.

In Fig. 6, the stress σ is represented against the crack growth, $2(a - a_o)$. At the first steps, the stress increases while the crack grows. Then, after reaching a maximum, the stress decreases and attains the value zero when the whole ligament is separated. The maximum represents the transition

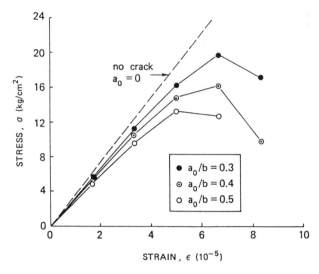

FIG. 5. Stress versus strain response for three different initial crack lengths.

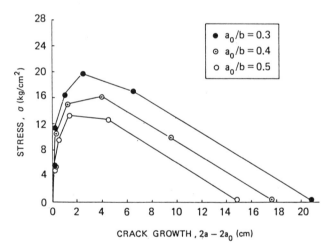

FIG. 6. Stress versus crack growth plots for three different initial crack lengths.

between stable and unstable structural behaviour. On the other hand, the transition between stable and unstable crack propagation depends on the achievement of the critical value of the strain energy density factor, S_c. Such a value has physical dimensions different from those of $(dW/dV)_c$, and produces the scale effects recurrent in Fracture Mechanics. In the reality, the crack instability may precede or follow the structural instability. This mainly depends on the structural size scale.

Because material damage and crack growth occur in a non-self-similar fashion for each step of loading, specimens of different sizes appear to behave differently. The application of Buckingham's theorem for physical similitude and scale modelling gives:

$$\sigma = \Pi \left[\varepsilon, \frac{E}{\left(\dfrac{\mathrm{d}W}{\mathrm{d}V}\right)_{\mathrm{c}}}, \frac{\sigma_{\mathrm{u}}}{\left(\dfrac{\mathrm{d}W}{\mathrm{d}V}\right)_{\mathrm{c}}}, v, \frac{l}{b}, \frac{t}{b}, \frac{a_{\mathrm{o}}}{b} \right] \tag{14}$$

where material toughness, $(\mathrm{d}W/\mathrm{d}V)_{\mathrm{c}}$, and specimen width, b, have been used as the fundamental quantities. The stress, σ, may be regarded as a function of the strain ε only, if all other ratios are kept constant.

In the same way, it is possible to define a dimensionless strain energy density factor:

$$\frac{S}{\left(\dfrac{\mathrm{d}W}{\mathrm{d}V}\right)_{\mathrm{c}} b} = \Sigma \left[\frac{a}{b}, \frac{E}{\left(\dfrac{\mathrm{d}W}{\mathrm{d}V}\right)_{\mathrm{c}}}, \frac{\sigma_{\mathrm{u}}}{\left(\dfrac{\mathrm{d}W}{\mathrm{d}V}\right)_{\mathrm{c}}}, v, \frac{l}{b}, \frac{t}{b}, \frac{a_{\mathrm{o}}}{b} \right] \tag{15}$$

Function Σ can be regarded as linear in a/b^{6}

$$\frac{S}{\left(\dfrac{\mathrm{d}W}{\mathrm{d}V}\right)_{\mathrm{c}} b} = \frac{\mathrm{d}S/\mathrm{d}a}{(\mathrm{d}W/\mathrm{d}V)_{\mathrm{c}}} \frac{a - a_{\mathrm{o}}}{b} + \frac{S_{\mathrm{o}}}{\left(\dfrac{\mathrm{d}W}{\mathrm{d}V}\right)_{\mathrm{c}} b} \tag{16}$$

which may obviously be rearranged into the form:

$$\frac{S}{\left(\dfrac{\mathrm{d}W}{\mathrm{d}V}\right)_{\mathrm{c}} b} = A \left(\frac{a}{b} \right) + B \tag{17}$$

The constants A and B are dimensionless and scale independent. It follows that the slope of the S–a diagram is constant varying the scale and the intercept, S_{o}, is proportional to the scale b.

Figure 7 shows the straight line plots of S versus crack growth for increasing size $b\,(a_{\mathrm{o}}/b = 0.3)$. The critical crack growth decreases with increasing specimen size. For example, with the critical value $S_{\mathrm{c}} = 8 \times 10^{-3}$ kg/cm, the limiting size is $b = 240$ cm. Beyond this size, stable crack

growth ceases to occur and failure corresponds to unstable crack propagation or catastrophic fracture.

Figure 8 presents the relations between stress and strain ($a_0/b = 0.3$). The
vertical lines with arrows indicate the limiting values of ε as the critical

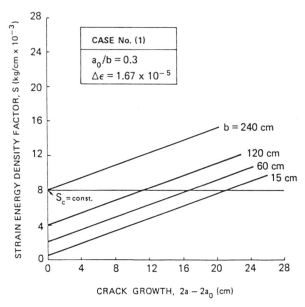

FIG. 7. Strain energy density factor versus crack growth plots by varying size b ($a_0/b = 0.3$).

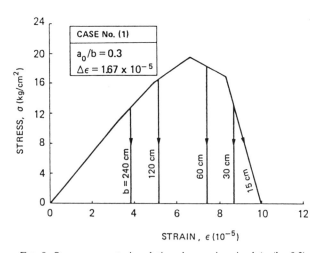

FIG. 8. Stress versus strain relations by varying size b ($a_0/b = 0.3$).

strain energy density factor, $S_c = 8 \times 10^{-3}$ kg/cm, is reached. Crack instability occurs for smaller strains as the size b increases. This is obvious, since the initial crack length a_o also increases, the ratio $a_o/b = 0.3$ being constant. The structural instability occurs before crack instability only for $b \lesssim 80$ cm (Fig. 8). When, for example, $b = 120$ cm, softening behaviour is not present and the crack starts to spread in an unstable manner, stress σ still being in the ascending stage.

It is also interesting to consider the maximum stress $\sigma_{max}^{(2)}$ resulting from the Linear Elastic Fracture Mechanics (LEFM) solution, eqn. (4), and $\sigma_{max}^{(3)}$ coming from the limit analysis at the ligament, eqn. (5). Normalizing the strengths $\sigma_{max}^{(1)}$ and $\sigma_{max}^{(3)}$, obtained respectively by the present approach and by the limit analysis, with $\sigma_{max}^{(2)}$, obtained through LEFM, we can evaluate the interaction between the different failure modes (Fig. 9). The

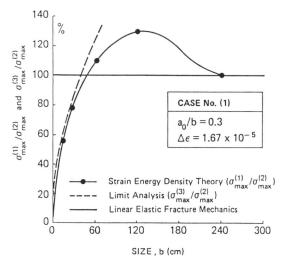

FIG. 9. Transition between plastic collapse and brittle crack propagation by varying size b $(a_o/b = 0.3)$.

horizontal line $\sigma_{max}^{(1)}/\sigma_{max}^{(2)}$ and $\sigma_{max}^{(3)}/\sigma_{max}^{(2)}$ equal to 100% represents the case when failure coincides totally with *brittle fracture*, while the dashed line represents the case when failure coincides totally with *plastic collapse*. When the specimen size is small, the simple formula in eqn. (5) gives good prediction based on the *ultimate strength* alone. On the other hand, when the specimen size is large, eqn. (4) gives good prediction based on the *stress-intensity factor* alone. The two extreme situations are then connected by a transition. For intermediate sizes, the ratio of Fig. 9 appears higher than one. This means that a fictitious critical stress-intensity factor $K_{IC}^{(f)}$ larger than the true K_{IC} may be assumed. In this way, the energy-absorbing damage process at the crack tip is taken into account.

4. SIZE EFFECTS ON STRENGTH, TOUGHNESS AND DUCTILITY

Equation (16) may obviously be rearranged into the form:

$$\tilde{S} = \tilde{S}'(\xi - \xi_0) + \tilde{S}_0 \tag{18}$$

the constants $\tilde{S}'$ and $\tilde{S}_0$ being dimensionless and scale independent. On the other hand, it is apparent that the quantity $S_c/(dW/dV)_c b$ must also enter into the dimensional analysis in eqn. (14). In fact, for estimating σ_{max}, it suffices to consider:

$$\frac{\sigma_{max}}{\left(\dfrac{dW}{dV}\right)_c b^2} = \pi(S^*) \tag{19}$$

in which S^* is a brittleness number analogous to that defined in eqn. (3):

$$S^* = \frac{S_c}{\left(\dfrac{dW}{dV}\right)_c b} = \frac{r_c}{b} \tag{20}$$

Hence, all geometrically similar structures can be regarded as governed by S^*. This dimensionless quantity can be used to predict the stress versus strain behaviour for all specimen sizes.[6]

In conclusion and recalling eqn. (18), it is possible to state that above the size (Fig. 10):

$$b_{max} = \frac{S_c}{\left(\dfrac{dW}{dV}\right)_c \tilde{S}_0} = \frac{r_c}{\tilde{S}_0} \tag{21}$$

stable crack growth ceases to occur and the brittle failure is achieved when the σ–ε curve is still in its linear initial course (Fig. 8), whereas below the size (Fig. 10):

$$b_{min} = \frac{S_c}{\left(\dfrac{dW}{dV}\right)_c [\tilde{S}_0 + \tilde{S}'(1 - \xi_0)]} = \frac{r_c}{\tilde{S}_0 + \tilde{S}'(1 - \xi_0)} \tag{22}$$

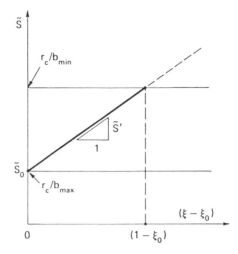

FIG. 10. Scale invariance of the non-dimensional crack growth resistance curve.

unstable crack growth ceases to occur (even in the softening stage) and the progressive slow crack growth develops up to the complete specimen separation. On the other hand, for $b_{min} < b < b_{max}$, stable (or slow) crack growth is followed by unstable (or fast) crack propagation. Ductility appears therefore as a mechanical property depending on the size scale of the structure, as well as on the fracture toughness of the material.

5. COOLING-HEATING EFFECT IN A HETEROGENEOUS MATERIAL SUBJECTED TO REPEATED LOADING

5.1. Description of Material and Testing Procedure

In spite of the apparent variation of the constitutive behaviour as a function of the size of the material element, the damage mechanics is proposed as an invariant process at the microscale, which can be revealed experimentally by temperature fluctuations.

The first of a series of tests on composite materials, such as concrete, is described within the framework of an investigation aimed at identifying a thermodynamic model for predicting the fatigue life of these materials when subjected to repeated compression loading. The tests reveal a marked 'cooling-heating' effect for very modest maximum compressive stresses ($\sigma_{max} = \frac{1}{5} - \frac{1}{10} f_c$, where f_c is the ultimate compressive strength). The repeated loading cycles produce a decrease in the temperature of the material at first, and then, as the applied load increases, the temperature is seen to increase. The behaviour is consistent with the findings of recent investigations on steel and aluminium specimens tested in tension.[10-13] However, since we are dealing with compression tests and repeated loading, a further investigation was considered useful.

It is verified that the 'cooling-heating' effect depends on the value of the maximum applied stress (σ_{max}). The test specimens subjected to stress $\sigma_{max} \gtrsim 0.5 f_c$ present the heating stage only (in this case the cooling phase may occur and yet be concealed). In the experimental program, carried out at the Materials Testing Laboratory of the Structural Engineering Department of the Politecnico di Torino, another material was also considered (plexiglass) for comparison purposes.

The first material is concrete made with 350 kg/m^3 Portland cement and alluvial aggregate with max. diameter $\phi = 8 \text{ mm} = 0.31$ in. The compressive strength was found to be $f_c = 37$ MPa. The test is performed on a cylindrical specimen ($H = D = 2.8 \text{ cm} = 1.10$ in), thermally insulated from the environment by means of polystyrene applied to end and side faces (Fig. 11). The temperature of the specimen is measured by means of two thermocouples, T_1 and T_2, applied to the side faces, and a third thermocouple T_3 is used to determine the temperature of the environment.

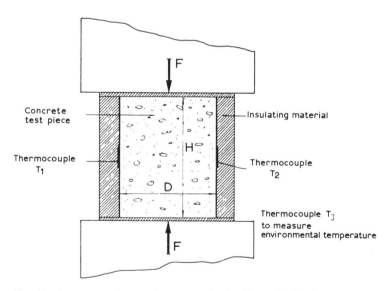

FIG. 11. Concrete specimen under compression loading cycles. Testing apparatus.

TABLE 1
Time–temperature values for a composite material subjected to compression fatigue testing

Type of load and application frequency	Time (s)	T_1 θ_1 (°C)	T_2 θ_2 (°C)	T_3 θ_3 (°C)	$\Delta\theta = \dfrac{\theta_1+\theta_2}{2} - \theta_3$ (K)(°C)
Load I	0	18·3	18·3	18·3	0
	60	18·3	18·0	18·3	−0·15
	120	18·3	18·0	18·5	−0·35
$\sigma_{min} = 1$ MPa	180	18·3	18·0	18·5	−0·35
$\sigma_{max} = 5$ MPa	240	18·0	17·8	18·3	−0·40
Frequency = 2 Hz	300	17·8	17·8	18·3	−0·50
	360	17·5	17·5	18·3	−0·80
	420	17·5	17·5	18·3	−0·80
	480	17·3	17·5	18·3	−0·90
	540	17·3	17·3	18·3	−1·00
	600	17·3	17·3	18·3	−1·00
Load II	660	17·3	17·3	18·3	−1·00
	720	17·3	17·3	18·3	−1·00
$\sigma_{min} = 1$ MPa	780	17·3	17·3	18·3	−1·00
$\sigma_{max} = 10$ MPa	840	17·3	17·3	18·3	−1·00
Frequency = 2 Hz	900	17·3	17·3	18·3	−1·00
	960	17·3	17·5	18·3	−1·00
Load III	1 020	17·5	17·8	18·5	−0·85
	1 080	17·8	17·8	18·5	−0·70
$\sigma_{min} = 1$ MPa	1 140	18·0	18·0	18·5	−0·50
$\sigma_{max} = 20$ MPa	1 200	18·0	18·0	18·5	−0·50
Frequency = 2 Hz	1 260	18·3	18·3	18·5	−0·20
	1 320	18·3	18·3	18·5	−0·20
	1 380	18·3	18·3	18·5	−0·20
	1 440	18·3	18·3	18·5	−0·20
Load III(a)	1 500	18·7	18·5	18·7	−0·10
	1 560	18·7	18·7	18·7	0·00
$\sigma_{min} = 1$ MPa	1 620	19·0	19·0	18·7	+0·30
$\sigma_{max} = 20$ MPa	1 680	19·2	19·2	18·7	+0·50
Frequency = 6 Hz					
Load IV	1 740	19·2	19·2	18·7	+0·50
	1 800	19·2	19·2	18·7	+0·50
$\sigma_{min} = 1$ MPa	1 860	19·2	19·2	18·7	+0·50
$\sigma_{max} = 30$ MPa	1 860		test piece failure		
Frequency 2 Hz					

The comparative specimen of plexiglass consists of a prism $(3 \times 3 \times 5$ cm) tested under identical conditions.

The temperature versus time curves are plotted, the specimens being subjected to repeated compression loading by means of a MTS machine (max. loading capacity of 100 000 N) and by measuring the temperature values by means of a Compact Logger 3440 with a sensitivity of 0·1 K.

5.2. Results and Comments

Table 1 reports the concrete specimen temperature values in relation to load variations. The load is characterized by the σ_{min} and σ_{max} values and the oscillation frequency.

The temperature variation $\Delta T (\Delta T = T_i - T_e) =$ specimen temperature – environmental temperature $= ((T_1 + T_2)/2) - T_3$ is plotted in Fig. 12.

For load I($\sigma_{max}/f_c = 0\cdot135$) at an oscillation frequency of 2 Hz we observe a marked drop in temperature (cooling stage) which becomes stable around an asymptotic value $\Delta T \simeq 1$ K. This value remains unchanged with load II ($\sigma_{max}/f_c = 0\cdot27$). A reversal in temperature is seen to occur when load

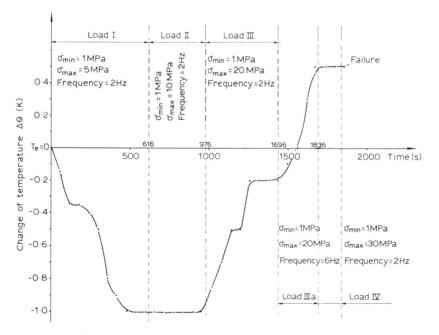

FIG. 12. Temperature versus time diagram for a concrete specimen under compression loading cycles at frequencies = 2–6 Hz.

III is applied ($\sigma_{max}/f_c = 0.54$). At this last stage, the effects of the load oscillation frequency on temperature values can be observed. The temperature, after reaching a plateau, increases when the frequency is raised from 2 to 6 Hz.

Figure 13 shows the results obtained for the plexiglass specimen. The

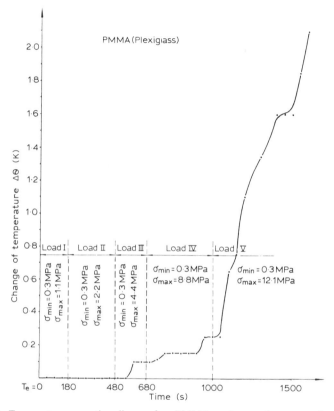

FIG. 13. Temperature versus time diagram for a PMMA specimen under compression loading cycles at a frequency of 2 Hz.

same testing procedure adopted for the concrete test was used. The cooling stage does not appear, at least according to the sensitivity of the equipment used ($\pm$ 0·1 K). Beginning with load III ($\sigma_{max} = 4.4$ MPa) a marked increase in temperature occurs.

It should be observed that the stress–strain curves obtained for plexiglass during the test by means of electrical strain gauges turn out to be perfectly overlapping with increasing cycle numbers. This means that energy dissipation can be neglected. The opposite happens in the case of concrete, for which the hysteretic curve shifts in the direction of positive strains, with permanent deformation.

The 'cooling-heating' effect may be explained in terms of the concepts of order and disorder formulated in the 'thermodynamics of irreversible processes'. The material's cooling coincides with the period in which order is established in the system with respect to the initial state. Physically, this could correspond to an increase in the material's homogeneity. In this way, the differences between concrete and plexiglass may be evidenced. In plexiglass, in fact, the homogenisation stage is much less pronounced than in concrete: therefore, cooling is not detected. The temperature versus time curve represents a possibility of measuring the state of order and disorder and therefore it becomes a precious tool to forecast the fatigue life of structural materials. The minimum in the temperature versus time diagram seems to represent the transition between order and disorder, from a thermodynamic point of view, or between reversible deformation and irreversible damage, from a mechanical point of view.

ACKNOWLEDGEMENT

The financial support provided by MPI is gratefully acknowledged by the authors.

REFERENCES

1. CARPINTERI, A., Static and energetic fracture parameters for rocks and concretes, *Materials & Structures (RILEM)*, **14** (1981), 151–62.
2. CARPINTERI, A., Notch sensitivity in fracture testing of aggregative materials, *Engineering Fracture Mechanics*, **16** (1982), 467–81.
3. CARPINTERI, A., Application of fracture mechanics to concrete structures, *Journal of the Structural Division (ASCE)*, **108** (1982), 833–48.
4. CARPINTERI, A., Plastic flow collapse vs. separation collapse in elastic–plastic strain-hardening structures, *Materials & Structures (RILEM)*, **16** (1983), 85–96.
5. CARPINTERI, A., Statistical strength variation in materials with a random distribution of defects, Nota Tecnica n. 73, Istituto di Scienza delle Costruzioni, Università di Bologna, 1983.
6. CARPINTERI, A. and SIH, G. C., Damage accumulation and crack growth in bilinear materials with softening, Report IFSM-83-115, Institute of Fracture and Solid Mechanics, Lehigh University, 1983.

7. SIH, G. C., Some basic problems in fracture mechanics and new concepts, *Engineering Fracture Mechanics*, **5** (1973), 365–77.

8. SIH, G. C. and MACDONALD, B., Fracture mechanics applied to engineering problems—Strain energy density fracture criterion, *Engineering Fracture Mechanics*, **6** (1974), 361–86.

9. HILTON, P. D., GIFFORD, L. N. and LOMACKY, O., Finite element fracture mechanics of two dimensional and axisymmetric elastic and elastic–plastic cracked structures, Report n. 4493, Naval Ship Research and Development Center, 1975.

10. BOTTANI, C. E. and CAGLIOTI, G., Mechanical instabilities of metals: temperature changes during deformation and indices of fundamental process, *Journal of Metallurgical Science and Technology*, **2** (1984), 3–7.

11. SIH, G. C. and TZOU, D. Y., Irreversibility and damage of SAFG-4OR steel specimen in uniaxial tension, *Theoretical Applied Fracture Mechanics*, **7** (1987), 23–30.

12. SIH, G. C., LIEU, F. L. and CHAO, C. K., Thermal/mechanical damage of 6061-T6 aluminium tensile specimen, *Theoretical Applied Fracture Mechanics*, **7** (1987), 67–78.

13. SIH, G. C., Mechanics and physics of energy density and rate of change of volume with surface, *Theoretical Applied Fracture Mechanics*, **4** (1985), 157–73.

9

Theoretical Modelling of Damage in Composite Laminates Subject to Low-Velocity Impact

J. WILLIAMS, I. H. MARSHALL

Paisley College of Technology, Paisley, UK

and

W. S. CARSWELL

National Engineering Laboratory, East Kilbride, UK

ABSTRACT

Herein is contained a theoretical assessment of the resulting damage in composite laminates subjected to low-velocity impact. The composite is considered to be planar isotropic as found in Chopped Strand Mat (CSM) and Sheet Moulding Compounds (SMC). The proposed theoretical methodology will, although approximate by necessity, be shown to reasonably predict the loss in strength and stiffness which occurs when such composites are subjected to impact damage. Reasonable comparison with experimental studies on glass/polyester CSM samples will be demonstrated.

NOTATION

W_s	Displacement of spherical impactor
W_p	Displacement of test plate
V_0	Impacting velocity
V_d	Volume of damage
$F(\tau)$	Force/time relationship
M	Indentor mass
F_1, F_2, etc.	Material failure stresses
F_m	Peak force during impact
E_j	Available energy to cause damage
T	Plate kinetic energy

133

U	Plate flexural energy
t	Maximum time being considered
τ	Any typical intermediate time, i.e. $0 \leqslant \tau \leqslant t$
m, n, i, j	Integer counters
a, b	Test specimen planform dimensions
c	Planform radius of indentation
q	Power index
w	Lateral plate displacement
h	Plate thickness
x, y, z	Plate cartesian co-ordinates
r, θ, z	Polar co-ordinates in damaged region
U_r, U_z	Local displacements at impacted region in the radial and transverse directions respectively
ρ	Mass density of composite
ω	Vibration frequency
α	Typical plate indentation
α_m	Maximum plate indentation
α_0	Permanent plate indentation
a^*, b^*	Characteristic crack dimensions
σ_{DS}	Damaged specimen residual strength
σ_{US}	Undamaged specimen residual strength
σ_0	Peak contact stress
σ_z	Transverse or through-thickness stress

1. INTRODUCTION

At the outset it should be stated that presently there exists no generalised means of analytically modelling the damage in composite structures subjected to impact damage. This is true for both ballistic and low-velocity impact damage. Moreover, the generic term given to studies of this nature, damage tolerance, is not uniquely defined and has a number of connotations depending on the particular field of study. For example, the term 'damage tolerance' has been used to describe the advanced stages of damage near the fatigue limit of structures subjected to cyclic loading. It has also been used to describe the ability of structures to successfully carry loading after localised impact damage, i.e. their tolerance to damage. An earlier paper by the authors[1] cited references reflecting a spectrum of usage of the term damage tolerance.

At this stage, it should be appreciated that the effects of low-velocity impact of composite materials will be appreciably different from the more traditional isotropic equivalents. An isotropic body will generally deform elastically, then plastically leaving permanent indentation, whereas the composite will undergo a multimode failure, i.e. the impacted energy will be transformed into matrix yielding and cracking, delamination, fibre cracking and pull-out or a complex mixture of these failure modes. Since each of these forms of failure is difficult to analytically model, their interaction produces extreme mathematical difficulties. Thus the whole concept of mathematically modelling failure mechanisms in composite structures is fraught with considerable difficulties irrespective of the available computational capability. Hence the reason for the present workshop which critically addresses this problem and assesses the present state of the art in this field.

2. THEORY

The methodology proposed herein will, of necessity, contain a number of simplifying assumptions. Although the same general philosophy can be applied to both thin and thick structures, the former will be specifically addressed as most composite structures are, relatively speaking, thin. Essentially the main difference in the response of thin and thick structures to low-velocity impact is flexural deformations which are prevalent in the former. Consequently, the energy consumed by such deformations must be accounted for when evaluating the available energy to cause damage of the composite.

Applying the Duhamel integral to the motion of the impactor, the plate equations (1) and (2) are obtained.

$$w_s = V_{0t} - \frac{1}{M} \int_0^t F(\tau)(t - \tau) \, d\tau \tag{1}$$

$$w_p = \frac{4}{ab\rho} \sum_m \sum_n \frac{1}{\omega_{mn}} \int_0^t F(\tau) \sin \omega_{mn}(t - \tau) \, d\tau \tag{2}$$

Hence the indentation of the test plate will be given by (3)

$$\alpha = V_0 t - \frac{1}{M} \int_0^t F(\tau)(t - \tau) \, d\tau - \frac{4}{ab\rho} \sum_m \sum_n \frac{1}{\omega_{mn}} \int_0^t F(\tau) \sin \omega_{mn}(t - \tau) \, d\tau \tag{3}$$

The impact of a thick plate will yield the force/time and displacement/time relationships shown in Fig. 1, with the equivalent curves for

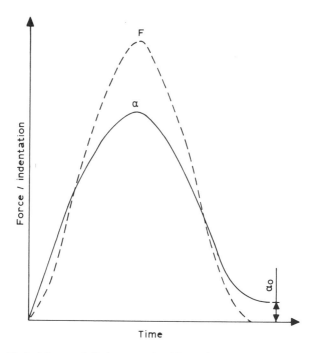

FIG. 1. Typical force and displacement time history for the impact of a thick plate.

a thin plate given in Fig. 2. Worthy of note is that eqn. (3) is only valid until the point of unloading, at which time the load–indentation law changes and eqn. (4), which is the general unloading law, can be used to represent the left hand side of eqn. (3). Typical repeated loading and unloading of a thin composite plate is shown in Fig. 3.

$$\alpha = \left[\frac{F(\tau)}{F_m}\right]^{1/q}(\alpha_m - \alpha_0) + \alpha_0 \tag{4}$$

This law was experimentally verified as diagrammatically illustrated in Fig. 4. By numerically solving eqns (3) and (4) by the mean force method, the complete loading history can be obtained.

The energy available to cause damage will not be totally absorbed by the plate and can be split into three main headings; that which causes damage, that which induces plate vibrations and the rebound kinetic energy. The energy to cause damage is given by eqn. (5). The vibrational energy comes from two parts, kinetic and bending, eqns (6) and (7), with the total given

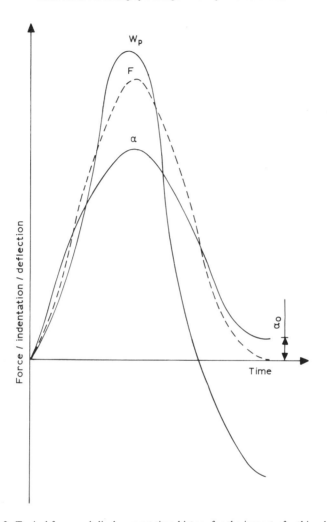

FIG. 2. Typical force and displacement time history for the impact of a thin plate.

by eqn. (8).

$$E_j = \int_0^t F_j(\omega_j - \omega_{j-1}) \tag{5}$$

$$T = \frac{1}{2}\rho h \iint \frac{\partial w^2}{\partial t} \, dx \, dy \tag{6}$$

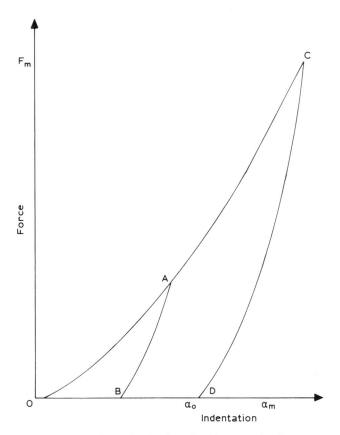

FIG. 3. Loading and unloading of a thin composite plate.

$$U = \frac{1}{2} \int\int \left[D_1 \left(\frac{\partial^2 w}{\partial x^2} \right)^2 + 2D_1 \mu_{21} \frac{\partial^2 w}{\partial x^2} \frac{\partial^2 w}{\partial y^2} + D_2 \left(\frac{\partial^2 w}{\partial y^2} \right)^2 \right.$$

$$\left. + 4D_k \left(\frac{\partial^2 w}{\partial x \partial y} \right)^2 \right] dx\, dy \tag{7}$$

$$E_{\text{Total}} = T + U \tag{8}$$

The stress/strain (9) and strain/displacement (10) relationships can be stated as:

$$[\sigma_i] = [a_{ij}]^{-1} [\varepsilon_j] \tag{9}$$

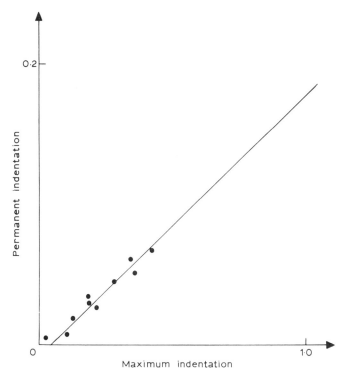

FIG. 4. Permanent indentation versus maximum indentation.

$$\varepsilon_r = \frac{\partial u_r}{\partial_r}$$

$$\varepsilon_\theta = \frac{u_r}{r}$$

$$\varepsilon_z = \frac{\partial u_z}{\partial z} \tag{10}$$

$$Y_{\theta z} = 0$$

$$Y_{rz} = \frac{\partial u_z}{\partial r} + \frac{\partial u_r}{\partial z}$$

$$Y_{r\theta} = 0$$

The internal displacements, eqn. (11) can be obtained by curve fitting a combination of experimental and finite element solution points.

$$U_z = \frac{\alpha_m}{\left(1 + \dfrac{z}{\alpha_m}\right)} \frac{4}{\left(4 + \left(\dfrac{r}{c}\right)^2\right)} \cos\left(\frac{\pi r}{3c}\right) \cos^4\left(\frac{\pi r}{6c}\right)$$

$$U_r = \frac{\alpha_m}{2} \frac{1}{\left[1 + \left(\dfrac{z}{\alpha_m}\right)^2\right]} \left(\frac{r^2}{3c^2 + r^2}\right)$$

(11)

Thus the internal stresses can be evaluated by substitution of (11) into (10) and (9).

The foregoing results are appropriate in the vicinity of the impacted region. Thereafter, the areas remote from damage can be considered using a cartesian set of co-ordinates.

The in-plane strains remote from the damaged region can be written as:

$$\varepsilon_x = -z \frac{\partial^2 w}{\partial x^2}$$

$$\varepsilon_y = -z \frac{\partial^2 w}{\partial y^2}$$

(12)

$$Y_{xy} = -2z \frac{\partial^2 w}{\partial x \partial y}$$

The plate stresses and strains can be related by eqn. (13).

$$\varepsilon_x = a_{11}\sigma_x + a_{12}\sigma_y + a_{16}\tau_{xy}$$

$$\varepsilon_y = a_{12}\sigma_x + a_{22}\sigma_y + a_{26}\tau_{xy}$$

(13)

$$\gamma_{xy} = a_{16}\sigma_x + a_{26}\sigma_y + a_{66}\tau_{xy}$$

Therefore, the in-plate stresses are given by eqn. (14).

$$\sigma_x = -z \left[A_{11}\frac{\partial^2 w}{\partial x^2} + A_{12}\frac{\partial^2 w}{\partial y^2} + 2A_{16}\frac{\partial^2 w}{\partial x \partial y} \right]$$

$$\sigma_y = -z \left[A_{12}\frac{\partial^2 w}{\partial x^2} + A_{22}\frac{\partial^2 w}{\partial y^2} + 2A_{26}\frac{\partial^2 w}{\partial x \partial y} \right]$$

(14)

$$\tau_{xy} = -z \left[A_{16}\frac{\partial^2 w}{\partial x^2} + A_{26}\frac{\partial^2 w}{\partial y^2} + 2A_{66}\frac{\partial^2 w}{\partial x \partial y} \right]$$

From simple equilibrium of a typical plate element eqns (15) can be formed.

$$\frac{\partial \sigma_x}{\partial x} + \frac{\partial \tau_{xy}}{\partial y} + \frac{\partial \tau_{xz}}{\partial z} = 0$$

$$\frac{\partial \tau_{xy}}{\partial x} + \frac{\partial \sigma_y}{\partial y} + \frac{\partial \tau_{yz}}{\partial z} = 0$$

(15)

Thus the corresponding through-thickness shear stresses are given by eqn. (16).

$$\tau_{xz} = \frac{1}{2}\left[z^2 - \frac{h^2}{4}\right]\left[A_{11}\frac{\partial^3 w}{\partial x^3} + 3A_{16}\frac{\partial^3 w}{\partial x^2 \partial y} + (A_{12} + 2A_{66})\frac{\partial^2 w}{\partial x \partial y^2} + A_{26}\frac{\partial^3 w}{\partial y^3}\right]$$

(16)

$$\tau_{yz} = \frac{1}{2}\left[z^2 - \frac{h^2}{4}\right]\left[A_{16}\frac{\partial^3 w}{\partial x^3} + 3A_{26}\frac{\partial^3 w}{\partial x \partial y^2} + (A_{12} + 2A_{66})\frac{\partial^2 w}{\partial x^2 \partial y} + A_{26}\frac{\partial^3 w}{\partial y^3}\right]$$

Considering the specific case of a rectangular planform plate being impacted by a spherical indentor, the total combined stresses can be given as eqn. (17), with the individual stresses transformed into the polar co-ordinate system by eqns (18).

$$\sigma_{\text{Total}} = \sigma_{\text{Local}} + \sigma_{\text{Bending}}$$

(17)

$$\sigma_r = \sigma_x \cos^2 \theta + \sigma_y \sin^2 \theta + 2\tau_{xy}\sin\theta\cos\theta$$
$$\sigma_\theta = \sigma_x \sin^2 \theta + \sigma_y \cos^2 \theta - 2\tau_{xy}\sin\theta\cos\theta$$
$$\sigma_z = \sigma_z$$
$$\tau_{\theta z} = -\tau_{xz}\sin\theta + \tau_{yz}\cos\theta$$
$$\tau_{rz} = -\tau_{xz}\cos\theta + \tau_{yz}\sin\theta$$
$$\tau_{r\theta} = (\sigma_y - \sigma_x)\sin\theta\cos\theta + \tau_{xy}(\cos^2\theta - \sin^2\theta)$$

(18)

With the internal stresses evaluated, the general stress distribution is given by Fig. 5, if a suitable failure criterion is applied on three mutually perpendicular faces of an element using eqns (19).

$$\frac{\sigma_1^2}{F_1^2} - \frac{\sigma_1 \sigma_2}{F_1 F_2} + \frac{\sigma_2^2}{F_2^2} + \frac{\sigma_{12}^2}{F_{12}^2} = 1$$

$$\frac{\sigma_2^2}{F_2^2} - \frac{\sigma_2 \sigma_3}{F_2 F_3} + \frac{\sigma_3^2}{F_3^2} + \frac{\sigma_{23}^2}{F_{23}^2} = 1$$

$$\frac{\sigma_1^2}{F_1^2} - \frac{\sigma_1 \sigma_3}{F_1 F_3} + \frac{\sigma_3^2}{F_3^2} + \frac{\sigma_{13}^2}{F_{13}^2} = 1$$

(19)

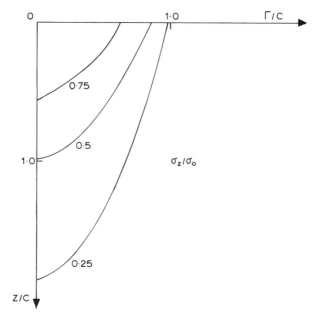

FIG. 5. Normal stress distribution.

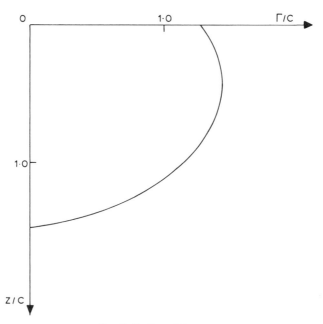

FIG. 6. Section of damage.

Using this methodology, a typical symmetrical section of the resulting volume of damage will be shown in Fig. 6. By representing the characteristic dimensions of the volume of damage using a semi-elliptical crack, as shown in Fig. 7, and making the resulting behaviour of the damaged and cracked specimens identical, a measure of the residual properties of the damaged specimen can be obtained.

The application of linear elastic fracture mechanics gives the strain energy release rate as eqn. (20).

$$G_I = \frac{(1-\mu^2)}{E}\pi K_I^2 \qquad (20)$$

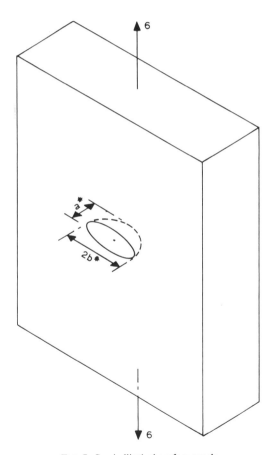

FIG. 7. Semi-elliptical surface crack.

where the stress intensity factor is given by eqn. (21).

$$K_{\mathrm{I}} = \left(1 + 0 \cdot 12\left(1 - \frac{a^*}{b^*}\right)\right)\sigma(\pi a^*)^{1/2}\,\frac{1}{\left(\dfrac{3\pi}{8} + \dfrac{\pi b^{*2}}{8a^{*2}}\right)}\left(\frac{2h}{\pi a^*}\tan\frac{\pi a^*}{2h}\right)^{1/2} \quad (21)$$

Thus the residual strength can be evaluated, eqn. (22).

$$\sigma_{\mathrm{DS}} = \sigma_{\mathrm{US}}\left[1 + \frac{1}{K_1 W_{\mathrm{us}}}[w_{\mathrm{ke}} - F(v_{\mathrm{d}})]\right]^{1/2} \quad (22)$$

Providing a relationship between stress and strain is available, the residual stiffness can also be evaluated.

3. RESIDUAL STRENGTH INDICATIONS

Typical residual strengths for thick and thin plates are shown in Figs. 8 and 9, respectively. The latter figure shows a number of experimental points obtained using CSM-GRP specimens which corroborate the present

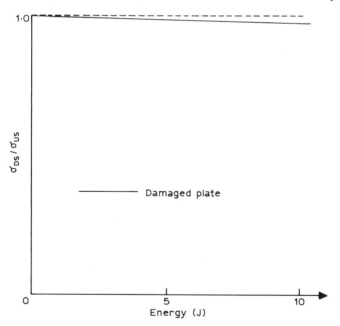

FIG. 8. Residual damaged strength (thick plates).

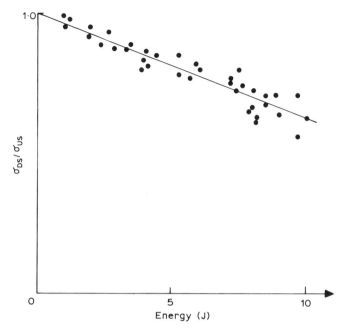

FIG. 9. Experimental and theoretical residual strength (thin plates).

theoretical methodology. Also, the former figure clearly shown that appreciably greater levels of damage energy are necessary to produce significant loss of strength in relatively thick specimens.

Clearly the present methodology has wide connotations with regard to various forms of composite materials, and indeed, traditional materials of construction, the only proviso being the assumption of planar isotropy.

REFERENCE

1. MARSHALL, I. H., WILLIAMS, J. and CARSWELL, W. S., Low velocity impact damage of CSM laminates, *Reinforced Plastics Congress 86*, British Plastics Federation Publ. 293/2, pp. 163–6.

10

Exact Elastic Stress Analysis
of Laminated Plates

A. J. M. SPENCER

Department of Theoretical Mechanics, University of Nottingham, UK

ABSTRACT

A forthcoming paper describes an exact method of solution of the three-dimensional elasticity equations for both the stretching and bending of isotropic laminated elastic plates. The theory is a generalisation to laminates of the 'exact plane stress' theory for homogeneous plates which was formulated by Michell in 1900 and was described by Love.[4] The essential feature is that any solution of the two-dimensional classical elastic plate equations can be used to generate an exact three-dimensional solution. The two-dimensional solution may be obtained by any of the available methods, including numerical methods. Among other quantities, the interfacial shear stress components are evaluated exactly.

Progress in extending this theory to the more important case of anisotropic laminates is described. In this case exact closed form solutions cannot be obtained, but solutions in infinite series of powers of the aspect ratio have been formulated.

This approach yields solutions which exactly satisfy the three-dimensional elasticity equations and lateral surface boundary conditions. Edge boundary conditions are satisfied only in an average sense. However the deviation of the edge boundary values given by the theory from those prescribed is a measure of the magnitude of the edge effects which may cause delamination and failure.

1. INTRODUCTION

The three-dimensional linear elastic stress distribution in a laminated plate is basic to any analysis of damage or failure in the plate. In this paper we outline a procedure for determining such stress distributions for plates of any shape subject to any of the standard edge boundary conditions. Both stretching and bending solutions may be dealt with, but for brevity we concentrate on the problem of in-plane stretching under edge loading. With the exception of a few very special solutions, the existing theories for elastic analysis of laminated plates are approximate theories, and none of them appears to the author to be entirely satisfactory. A brief review is given in Ref. 1. The most commonly used theory is probably the one known as Classical Laminate Theory, which is described, by e.g. Christensen.[2] In effect, in this theory the laminated plate is replaced by a homogeneous equivalent plate of the same overall geometry but with elastic constants which are suitably weighted averages of the elastic constants of the laminae. By its nature, such a theory can only yield averaged displacement components and approximate stress components. Whilst this knowledge is adequate for many purposes, for the analysis of failure it is important to know the detailed through-thickness distribution of stress and displacement so as to determine, for example, the inter-ply shear tractions which have an important bearing on the onset of delamination.

In a recent paper, Kaprielian et al.[1] have given a three-dimensional analysis for a laminate composed of *isotropic* laminae. This theory is outlined in Section 2. The outstanding feature of the theory is that the three-dimensional solution for the laminate is generated, in a very simple manner, by the two-dimensional classical laminate theory solution for the equivalent plate.

The case of greater practical interest is that in which the laminae are *anisotropic* and differently oriented. The theory of Section 2 does not extend directly to the case of anisotropic laminae, but it can be adapted to this case, to any required degree of accuracy. Work on the problem of anisotropic laminae is still incomplete, but a preliminary report on it is given in Section 3. It is intended that a full account will be published elsewhere.

The only restriction on these theories is that, as in all plate theories, edge boundary conditions can be satisfied only in an average fashion, rather than point by point. A brief discussion of edge effects is presented in Section 4.

2. ISOTROPIC LAMINATES

This section gives a brief summary of the theory which is described in detail in Ref. 1. We first state some results of classical isotropic thin plate theory. Our plate is supposed to lie in the X, Y plane of a system of rectangular

Cartesian coordinates X, Y, Z. Relative to this system, displacement components are denoted by U, V, W, and stress components by $\sigma_{xx}, \sigma_{xy}, \ldots,$ σ_{zz}. Then for in-plate stretching deformations the governing equations are

$$2(\eta+1)\Delta_{,X} - \Omega_{,Y} = 0, \quad 2(\eta+1)\Delta_{,Y} + \Omega_{,X} = 0 \tag{2.1}$$

where

$$\Delta = \bar{U}_{,X} + \bar{V}_{,Y}, \quad \Omega = \bar{V}_{,X} - \bar{U}_{,Y} \tag{2.2}$$

and

$$\eta = \lambda/(\lambda+2\mu) = \nu/(1-\nu) \tag{2.3}$$

where $\bar{U}$, $\bar{V}$ are *average* in-plane displacements, λ, μ are the Lamé elastic constants, ν is Poisson's ratio, and subscripts following commas denote partial derivatives. It follows from (2.1) that

$$\nabla^2 \Delta = 0, \qquad \nabla^2 \Omega = 0 \tag{2.4}$$

where ∇^2 is the two-dimensional Laplacian operator. For bending deformations under edge loading, the classical result is

$$\nabla^4 \bar{W} = 0 \tag{2.5}$$

where $\bar{W}$ is the deflection of the mid-plane. When $\bar{U}$ and $\bar{V}$, or $\bar{W}$, are determined, the corresponding stress resultants and moments are readily calculated.

We now present two classes of exact solutions in three-dimensional isotropic elasticity theory. We consider a homogeneous layer of uniform thickness $2h$ with its mid-plane at $Z = 0$. It is convenient to employ scaled variables (x, y, z) and (u, v, w) defined as

$$(x, y, z) = (X/a, Y/a, Z/a), \quad (u, v, w) = (U/a, V/a, W/a) \tag{2.6}$$

where a is a characteristic in-plane dimension. We also define the aspect ratio ε as

$$\varepsilon = h/a \tag{2.7}$$

In practice it is expected that normally $\varepsilon \ll 1$, but the analysis is exact and does not depend on ε being small.

Solution 1

Suppose that $u_0(x, y)$, $v_0(x, y)$ satisfy the classical thin plate equations for some (unspecified) value $\hat{\eta}$ of η, and

$$\Delta_0 = u_{0,x} + v_{0,y}, \quad \Omega_0 = v_{0,x} - u_{0,y} \tag{2.8}$$

Thus

$$2(\hat{\eta}+1)\Delta_{0,x}-\Omega_{0,y}=0, \quad 2(\hat{\eta}+1)\Delta_{0,y}+\Omega_{0,x}=0 \tag{2.9}$$

and so

$$\nabla^2\Delta_0=0, \quad \nabla^2\Omega_0=0 \tag{2.10}$$

Then the three-dimensional elasticity equations in the layer with elastic constant η are satisfied by $u(x, y, z)$, $v(x, y, z)$, $w(x, y, z)$, where

$$\begin{bmatrix} u(x, y, z) \\ v(x, y, z) \end{bmatrix} = \begin{bmatrix} u_0(x, y) \\ v_0(x, y) \end{bmatrix} - \varepsilon^2(\tfrac{1}{2}z^2 + S_2 z + S_3)\left\{(\eta+2)\begin{bmatrix}\Delta_{0,x}\\\Delta_{0,y}\end{bmatrix} + \begin{bmatrix}-\Omega_{0,y}\\\Omega_{0,x}\end{bmatrix}\right\}$$

$$\tag{2.11}$$

$$w(x, y, z) = -\varepsilon\eta(z+S_1)\Delta_0$$

where S_1, S_2, S_3 are arbitrary constants.

Solution 2

Suppose that $w_0(x, y)$ satisfies the biharmonic equation

$$\nabla^4 w_0 = 0 \tag{2.12}$$

Then a solution of the three-dimensional elasticity equations is

$$\begin{bmatrix} u(x, y, z) \\ v(x, y, z) \end{bmatrix} = -\varepsilon(z+B_1)\begin{bmatrix}w_{0,x}\\w_{0,y}\end{bmatrix} + \varepsilon^3(\tfrac{1}{6}z^3 + \tfrac{1}{2}B_1 z^2 + B_3 z + B_4)(\eta+2)\begin{bmatrix}\nabla^2 w_{0,x}\\\nabla^2 w_{0,y}\end{bmatrix}$$

$$\tag{2.13}$$

$$w(x, y, z) = w_0(x, y) + \varepsilon^2\eta(\tfrac{1}{2}z^2 + B_1 z + B_2)\nabla^2 w_0$$

where B_1 to B_4 are arbitrary constants.

For either solution, the stress is readily calculated by substituting (2.11) or (2.13) into the stress–strain relations.

The solutions (2.11) and (2.13) can be derived in various ways; one derivation is given in Ref. 1. With special choices of the constants S_i, B_i they were obtained by Michell[3] and are given in Love.[4] However these authors do not seem to have recognised that the three-dimensional solutions are generated by two-dimensional solutions of (2.9) and (2.12).

Let us now consider a laminate comprising $2N+1$ layers of different isotropic materials. For simplicity we consider geometrically symmetric laminates. Quantities relating to the rth layer will be identified by the index r; the layer $r=0$ contains the mid-plane of the laminate and the layer $r=N$ is adjacent to the upper surface.

Solutions of the form (2.11) and (2.13) are adopted in each layer, with the constants $S_i^{(r)}$ and $B_i^{(r)}$ applying in the rth layer. The solutions (2.11) are

appropriate for stretching deformations and (2.13) for bending deformations; for a symmetric laminate these solutions uncouple and may be treated separately. For brevity we discuss the stretching solution only. This has available the $3(N+1)$ constants $S_i^{(r)}$. Conditions which have to be satisfied are

(a) symmetry conditions at the mid-plane $z = 0$;
(b) continuity of displacement and traction at each interlaminar interface;
(c) traction-free conditions at the lateral surfaces.

It is found that all of these conditions can be satisfied by appropriate choices of $S_i^{(r)}$, $u_0(x, y)$ and $v_0(x, y)$. Conditions (a) and (b) provide a set of recurrence relations to determine the $S_i^{(r)}$; except for a single normalising condition these constants depend only on the laminate stacking geometry and laminae elastic constants. Hence for a given laminate they can be calculated once and for all, and do not depend on any particular boundary-value problem or shape of plate. The displacements u_0 and v_0 have to be solutions of the thin plate equations (2.9) with the elastic constant $\hat{\eta}$ chosen to have the value it has in the classical laminate theory—that is, the value appropriate for the equivalent plate.

We may therefore follow a very simple procedure. We first calculate the elastic constants for the equivalent plate, as in classical laminate theory, and the laminate constants $S_i^{(r)}$. Then, for a given boundary value problem, we solve the classical laminate theory equations to determine u_0 and v_0; this may be done by any available method, including numerical methods. The displacement in each layer is then given immediately by (2.11), and the stress by substitution in the stress–strain relations. The bending problem may be treated similarly.

One of the quantities of interest is the inter-laminar shear stress. If the traction on the interface between layers $r-1$ and r is τ_r, then we find that

$$\tau_r = -2\left\{ \mu_0\varepsilon_0(\eta_0 - \hat{\eta}) + 2\sum_{s=1}^{r-1} \varepsilon_s\mu_s(\eta_s - \hat{\eta}) \right\} \text{grad } \Delta_0 \qquad (2.14)$$

where μ_s, η_s relate to layer s, $\hat{\eta}$ relates to the equivalent plate, $\varepsilon_s = h_s/a$, $2h_s$ is the thickness of the layer s, and Δ_0 arises from the equivalent plate solution.

3. ANISOTROPIC LAMINAE

The success of the procedure described in Section 2 arises because the right-hand sides of (2.11) and (2.13), regarded as power series in z, terminate after a finite number of terms, and reduce to polynomials in z. When the laminae are anisotropic, this situation no longer obtains, and corresponding closed form solutions do not exist. However the concept of generating three-dimensional elasticity solutions from two-dimensional equivalent plate solutions remains valid.

We consider stretching deformations only. In classical thin plate theory, the governing equations for the mean in-plane displacement components are

$$
\begin{aligned}
&Q_{11}\bar{u}_{,xx} + 2Q_{16}\bar{u}_{,xy} + Q_{66}\bar{u}_{,yy} + Q_{16}\bar{v}_{,xx} \\
&+ (Q_{12} + Q_{66})\bar{v}_{,xy} + Q_{26}\bar{v}_{,yy} = 0 \\[2mm]
&Q_{16}\bar{u}_{,xx} + (Q_{12} + Q_{66})\bar{u}_{,xy} + Q_{26}\bar{u}_{,yy} \\
&+ Q_{66}\bar{v}_{,xx} + 2Q_{26}\bar{v}_{,xy} + Q_{22}\bar{v}_{,yy} = 0
\end{aligned}
\tag{3.1}
$$

where Q_{ij} are the 'reduced' elastic constants of the material. Solutions of these equations can be represented in the form (e.g. Leknitskii[5])

$$
\begin{bmatrix} \bar{u} \\ \bar{v} \end{bmatrix} = 2\,\mathrm{Re}\,\{\mathbf{p}_1\varphi_1(x + s_1 y) + \mathbf{p}_2\varphi_2(x + s_2 y)\}
\tag{3.2}
$$

where s_i and $\mathbf{p}_i$ are the eigenvalues and eigenvectors of the eigenvalue problem which arises by seeking solutions of (3.1) of the form $\bar{\mathbf{u}} = \mathbf{p}\varphi(x + sy)$.

For an anisotropic homogeneous layer we seek three-dimensional solutions of the form

$$
\mathbf{u} = \mathbf{U}\varphi\{x + sy + \varepsilon t(z - d)\}
\tag{3.3}
$$

where $\mathbf{u} = (u, v, w)^T$. On substituting (3.3) into the stress–strain relations and the equilibrium equations, and provisionally treating s as known, there results an eigenvalue problem with eigenvalues $\pm t_i$ and corresponding eigenvectors $\mathbf{U}_i^{\pm}$ ($i = 1, 2, 3$). Thus we obtain solutions for the rth layer of the form

$$
\begin{aligned}
\mathbf{u} = \sum_{i=1}^{3} \big[&K_i^{(r)+}\mathbf{U}_i^{(r)+}\varphi\{x + sy + \varepsilon_r t_i^{(r)}(z - d_i^{(r)+})\} \\
&+ K_i^{(r)-}\mathbf{U}_i^{(r)-}\varphi\{x + sy - \varepsilon_r t_i^{(r)}(z - d_i^{(r)-})\} \big]
\end{aligned}
\tag{3.4}
$$

where $K_i^{(r)\pm}$ and $d_i^{(r)\pm}$ are constants.

For a laminate we assume solutions of the form (3.4) in each layer. The constants $d_r^{\pm}$ are chosen so that the arguments of φ are continuous at each interface. The conditions of continuity of displacement and traction at the interfaces then determine $K_i^{(r)\pm}$ to leading order in ε except for normalising factors. It remains to determine the parameter s. This is fixed by the traction-free boundary conditions on the lateral surfaces of the plate. It is found that these are satisfied to leading order in ε if s is chosen to have the value given by classical plate theory, for the equivalent plate. Thus, we obtain solutions of the form

$$
\begin{aligned}
\mathbf{u} = \mathrm{Re} \sum_{i=1}^{3} \sum_{j=1}^{2} & [K_{ij}^{(r)+} \mathbf{U}_{ij}^{(r)+} \varphi_j \{x + s_j y + \varepsilon_r t_{ij}^{(r)}(z - d_i^{(r)+})\} \\
& + K_{ij}^{(r)-} \mathbf{U}_{ij}^{(r)-} \varphi_j \{x + s_j y - \varepsilon_r t_{ij}^{(r)}(z - d_i^{(r)-})\}]
\end{aligned}
\tag{3.5}
$$

where s_j are determined by the elastic constants for the equivalent plate, $\varphi_j(x + s_j y)$ represent a solution of classical laminate theory for the equivalent plate, and $t_{ij}^{(r)}$, $d_i^{(r)\pm}$, $K_{ij}^{(r)\pm}$, $\mathbf{U}_{ij}^{(r)\pm}$ can be determined in terms of the laminate geometry and the elastic constants of the laminae.

Solutions of the form (3.5) yield exact solutions of the linear elasticity equations in each lamina. However they satisfy the required continuity conditions at the inter-laminar interfaces and the traction-free boundary conditions at the lateral surfaces only to leading order in ε. To correct this we note that the discrepancy is of order $\varepsilon \varphi_j'(x + s_j y)$, and that if $\varphi_j(x + s_j y)$ generates a solution of the governing equations, so do $\varphi_j^{(n)}(x + s_j y)$, where $\varphi_j^{(n)}$ denotes the nth derivative. Thus, finally, we are led to solutions of the form

$$
\begin{aligned}
\mathbf{u} = \mathrm{Re} \sum_{i=1}^{3} \sum_{j=1}^{2} \sum_{n=0}^{\infty} & \varepsilon_r^n [K_{ijn}^{(r)+} \mathbf{U}_{ij}^{(r)+} \varphi_j^{(n)} \{x + s_j y + \varepsilon_r t_{ij}^{(r)}(z - d_i^{(r)+})\} \\
& + K_{ijn}^{(r)-} \mathbf{U}_{ij}^{(r-)} \varphi_j^{(n)} \{x + s_j y - \varepsilon_r t_{ij}^{(r)}(z - d_i^{(r)-})\}]
\end{aligned}
\tag{3.6}
$$

Thus, given a classical thin plate solution for the equivalent plate, the three-dimensional solution for the laminate can be constructed to any required order in ε.

4. EDGE EFFECTS

The solutions outlined in Sections 2 and 3, which we term interior solutions, are exact solutions of the appropriate elasticity equations, and satisfy the lateral surface boundary conditions. However they satisfy edge boundary conditions only in an average sense. To satisfy edge boundary conditions pointwise, it is necessary to superpose additional solutions which neutralise the mismatch between the actual boundary conditions and the boundary values of the interior solution. It is to be expected that these edge solutions will decay exponentially with distance from the edge; however, especially for highly anisotropic materials it is possible that the exponential decay rate will be slow and the edge effects may penetrate a substantial distance into the plate. Also, the edge solutions may include stress singularities at the intersection of an edge with an interlaminar interface.

Although the determination of the edge solution is a separate problem, the interior solution can give useful information about the magnitude of the stress in the edge solution. In particular, the magnitude of the mismatch between the actual boundary conditions and the boundary values of the outer solution controls the magnitude of the edge solution. It is tentatively suggested that some suitable measure of this mismatch might serve as a design parameter which could be used as a measure of susceptibility to edge delamination.

ACKNOWLEDGEMENTS

The research outlined in this paper was carried out in collaboration with Dr T. G. Rogers and Dr P. V. Kaprielian and full accounts are to be published jointly with these authors. The support of the Science and Engineering Research Council and of Rolls-Royce plc is gratefully acknowledged.

REFERENCES

1. KAPRIELIAN, P. V., ROGERS, T. G. and SPENCER, A. J. M. Theory of laminated elastic plates. I: Isotropic laminae, *Phil. Trans. Roy. Soc. A*, **324** (1988), 565–94.
2. CHRISTENSEN, R. M., *Mechanics of Composite Materials*, New York, John Wiley, 1979.
3. MICHELL, J. H., *Proc. Lond. Math. Soc.*, **31** (1900), 100–24.
4. LOVE, A. E. H., *The Mathematical Theory of Elasticity*, 4th edn, Cambridge, 1927.
5. LEKHNITSKII, S. G., *Theory of Elasticity of an Anisotropic Elastic Body*, San Francisco, Holden-Day, 1963.

11

Recent Progress in the Mathematical Modeling of Composite Materials

R. V. KOHN

Courant Institute of Mathematical Sciences, New York, USA

ABSTRACT

We review some of the recent mathematical progress on the effective moduli of composites. Specific attention is devoted to mathematically precise definitions of effective moduli, new methods for bounding effective moduli, new constructions of mixtures with explicitly computable properties, and applications to structural optimization.

1. INTRODUCTION

We are concerned with materials that are spatially heterogeneous on a suitably small length scale, and with linear models of material behavior, for example linear elasticity. The effective moduli of such a 'composite' describe its overall, large-scale behavior. They have long been an object of study by physicists and materials scientists; selective reviews of the extensive literature include Refs. 14, 22, 68, 69, 72. More recently, the study of effective moduli has attracted the attention of a growing community of mathematicians as well. The relatively new notions of homogenization and G-convergence provide a firm mathematical foundation;[48,62,65,71] moreover, the effective moduli of composites have been linked to fundamental issues arising in the optimal control of certain distributed parameter systems, and to deep questions involving the lower semicontinuity of variational functions, see, e.g. Refs. 1, 12, 28, 30, 32, 38, 39, 49, 50, 58, 67. The specific questions about effective moduli raised by these new applications are sometimes different from those that were the focus of the older literature: for example, applications to structural optimization require the specification of all (anisotropic) composites attainable as mixtures of given components in specified proportions. However, the mathematical tools developed to address such questions have also led to new results that are very much within the purview of the older theory. Examples include the simultaneous attainability of the Hashin–Shtrikman shear modulus and bulk modulus bounds;[17,36,42,51] the validity of a conjecture of Schulgasser about the effective conductivity of polycrystalline composites;[5] and the attainability of certain mean field theories.[2,41]

The goal of this paper is to review selected aspects of this recent mathematical progress, which it is hoped will be of interest to a broad community of specialists in materials science. It should be emphasized that the ideas presented here are a synthesis of the work of many individuals, and that the selection of topics is somewhat arbitrary—in no way representing a comprehensive survey of the most important recent developments.

2. MATHEMATICALLY PRECISE DEFINITIONS OF EFFECTIVE MODULI

We are concerned with mixtures of continua on a length scale small compared to that on which the loads and boundary conditions vary, but still large enough for continuum theory to apply. Such a 'composite' is clearly an idealization: it represents the limiting behavior of a sequence of structures, as the ratio $\varepsilon = l/L$ relating the 'microscopic' length scale l to the 'macroscopic' one L tends to zero. There are in fact several distinct theories, differing as to the form assumed for the fine scale structure. A *periodic* composite is one whose microscopic structure is periodic with a specified unit cell; a *random* composite is one whose fine scale structure is a stochastic process with specified statistics. There is also a third approach which makes no such hypothesis on the fine scale structure, appealing instead to a compactness theorem for systems of partial differential equations. This last theory, known variously as *G-convergence* or *homogenization*, represents in a sense the most general approach.

To fix ideas, let us focus the discussion on mixtures of two isotropic, linearly elastic materials in $\mathbf{R}^d$ ($d = 2$ and $d = 3$ being, of course, the cases of physical interest). Each of the component materials is characterized by a bulk modulus κ_i and a shear modulus μ_i ($i = 1,2$), determining a unique Hooke's law tensor A_i—a symmetric linear map on the space of symmetric tensors—such that

$$A_i e = \kappa_i (\text{tr } e) I + 2\mu_i \left(e - \frac{1}{d}(\text{tr } e) I \right) \qquad (2.1)$$

for any symmetric tensor e. The associated 'elastic energy' quadratic form is the inner product of stress and strain:

$$(A_i e, e) = \left(\kappa_i - \frac{2\mu_i}{d} \right)(\text{tr } e)^2 + 2\mu_i |e|^2 \qquad (2.2)$$

A structure which mixes the two materials will have a spatially varying Hooke's law, equal to either A_1 or A_2 at each material point x. Introducing a parameter ε, representing (at least in the periodic and random cases) the length scale of the microstructure, the spatially varying Hooke's law is

$$A^\varepsilon(x) = \chi_1^\varepsilon(x) A_1 + \chi_2^\varepsilon(x) A_2 \qquad (2.3)$$

where

$$\chi_i^\varepsilon(x) = \begin{cases} 1 & \text{on the set occupied by material } i \\ 0 & \text{elsewhere} \end{cases} \qquad (2.4)$$

so that $\chi_2^\varepsilon = 1 - \chi_1^\varepsilon$. By definition the structure is *periodic* (with cubic) symmetry) if

$\chi_i^\varepsilon(x) = \chi_i(\frac{x}{\varepsilon})$ for some function $\chi_i(y)$,
taking only the values 0 and 1, defined for all $y \in \mathbf{R}^d$
and periodic in each component of y with period 1. $\qquad (2.5)$

An example would be a periodic array of spherical inclusions centered on a cubic lattice of mesh ε, each sphere having radius $\varepsilon\rho$ ($\rho < \frac{1}{2}$). In the *random* case there is an additional variable ω, belonging to a suitable probability space:

$\chi_i^\varepsilon(x, \omega) = \chi_i(\frac{x}{\varepsilon}, \omega)$ for some stochastic process $\omega \to \chi_i(y, \omega)$,
defined for $y \in \mathbf{R}^d$ and ω in a probability space, and taking only the values 0 and 1. It is required that χ_i be translation invariant, in the sense that $\omega \to \chi_i(y + c, \omega)$ gives the same stochastic process for each $c \in \mathbf{R}^d$. Furthermore, the translations are assumed to be ergodic, so that ensemble averaging is equivalent to spatial averaging. $\qquad (2.6)$

An example would be a family of (possibly overlapping) spherical inclusions of radius $\varepsilon\rho$ whose centers have a multidimensional Poisson distribution, the expected number of balls in a unit-sized region being of order ε^{-d}. The hypotheses (2.5) or (2.6) specify rather precisely the character of the fine scale structure. The G-*convergence* approach, by contrast, makes no such hypothesis:

> $\chi_1^\varepsilon(x)$ is any family of functions taking only the values 0 and 1, parametrized by $\varepsilon \to 0$, and $\chi_2^\varepsilon = 1 - \chi_1^\varepsilon$. (2.7)

It is specifically *not* assumed in (2.7) that ε represents the length scale of the microstructure: even a sequence which has no clear separation of scales is permitted. Clearly (2.7) includes both the periodic case and the random one; indeed, in our opinion it includes any reasonable notion of a linearly elastic composite obtained by mixing two materials (with perfect bonding at all material interfaces).

The tensor of effective moduli A^* is simply the Hooke's law tensor of the composite. It represents the limiting behavior of the mixture as $\varepsilon \to 0$. This means that for any (ε-independent) load f, the associated elastostatic displacement u^ε—which solves the equilibrium equations

$$\sigma^\varepsilon = A^\varepsilon e^\varepsilon$$

$$e_{kl}^\varepsilon = \frac{1}{2}\left(\frac{\partial u_k^\varepsilon}{\partial x_l} + \frac{\partial u_l^\varepsilon}{\partial x_k}\right)$$ (2.8)

$$\operatorname{div} \sigma^\varepsilon = f$$

with an appropriate boundary condition—converges as $\varepsilon \to 0$ to u^*, the solution of the corresponding system with A^ε replaced by A^*. The starting point of the mathematical theory is the *existence* of effective moduli. In the spatially periodic and stationary stochastic contexts (2.5), (2.6), translation invariance assures that the tensor A^* of effective moduli is constant. For periodic composites it can be given in terms of the solutions of certain canonical 'cell problems', see for example, Refs. 8, 60, but we prefer this variational characterization, cf. Ref. 64:

$$(A^*\xi, \xi) = \inf_\phi \int_Q (\tilde{A}(y)[\xi + e(\phi)], \xi + e(\phi))\,dy$$ (2.9)

in which

$$\tilde{A}(y) = \chi_1(y)A_1 + \chi_2(y)A_2$$ (2.10)

$Q = [0, 1]^d$ is the unit cell of the perodic structure, ϕ varies over Q-periodic displacement fields, and $e(\phi) = \frac{1}{2}(\nabla\phi + \nabla\phi^T)$ is the linearized strain associated to ϕ. An entirely analogous formula is available in the random case, cf. Refs. 19, 33, 55, 70:

$$(A^*\xi, \xi) = \inf_{\mathbf{E}(e) = \xi} \mathbf{E}[(\tilde{A}e, e)] \tag{2.11}$$

in which $\mathbf{E}$ represents the ensemble average and e ranges over stationary, random strain fields with mean value ξ. In the more general G-convergence setting (2.7) there is no hypothesis of translation invariance, so the tensor of effective moduli $A^*(x)$ can vary with x. Moreover, there is obviously not enough structure to give a formula as explicit as (2.9) or (2.11). But it is nevertheless true that for any sequence χ_i^ε as in (2.7) there is a subsequence $\varepsilon' \to 0$ for which there exists a limiting tensor of effective moduli $A^*(x)$, see for example Refs. 48, 62, 65, 71.

We shall be interested in bounds for A^* in terms of the volume fractions of the component materials, so let us note here how to express these volume fractions in each of the different settings. For the periodic composite (2.5) the volume fraction of material i is the proportion of the period cell occupied by it:

$$\theta_i = \int_Q \chi_i(y)\, dy \tag{2.12}$$

Similarly, in the stationary, random case (2.6) it is the expected value of $\chi_i(y, \omega)$:

$$\theta_i = \mathbf{E}(\chi_i) \tag{2.13}$$

In the *G-convergence* context (2.7) it is instead given by the L^∞—weak* limit

$$\theta_i(x) = \text{wk}^* \lim_{\varepsilon \to 0} \chi_i^\varepsilon(x) \tag{2.14}$$

no longer necessarily constant, defined by the property that

$$\int \chi_i^\varepsilon(x) g(x)\, dx \to \int \theta_i(x) g(x)\, dx \tag{2.15}$$

for continuous functions g.

These notions of effective moduli are easily seen to be equivalent to the operational definitions more commonly used in materials science, based on the average stress and strain or average elastic energy in a physical domain that is large compared with the microstructure but small compared with the length scale of the loads and boundary conditions, see for example, Refs. 22, 24. They are important for the development of a proper mathematical theory, because they make it possible to give fully rigorous proofs of results about effective moduli. But why should they be of interest to a materials scientist? One answer lies in the following 'density' result:[16] if an algebraic relation between the tensor of effective moduli and the component volume fractions holds for all spatially periodic composites (or for all stationary, stochastic composites), then it holds in the more general context of G-convergence as well. Thus, *for bounds on effective moduli in terms of volume fractions alone, neither long-range disorder nor a definite separation of scales is relevant.* This resolves a point which has been the object of considerable controversy in the literature, see for example, Ref. 22.

3. NEW METHODS FOR BOUNDING EFFECTIVE MODULI

A typical goal of the new mathematical theory is the so-called *G-closure problem*: find the precise set of Hooke's laws A^* achievable by mixing two given isotropic, elastic materials in specified proportions. The motivation comes from applications to structural optimization, as we shall explain in Section 5. The special case when A^* is isotropic was considered by Hashin and Shtrikman,[23] under the further hypothesis that the component materials are well-ordered i.e. that

$$\mu_1 \leqslant \mu_2, \quad \kappa_1 \leqslant \kappa_2 \tag{3.1}$$

They gave upper and lower bounds for the effective bulk and shear moduli, κ^* and μ^*, which are now known to be simultaneously achievable.[17,51] An improvement of the Hashin–Shtrikman bounds can be found in Refs. 10 and 47, but the precise set of attainable isotropic composites is still not known. In any event, results of this kind—concerning A^* with specified symmetry—are not adequate for applications to structural optimization, since the best composites for use in an optimal structure can (and generally will) be fully anisotropic. While the complete solution of the G-closure problem seems beyond the reach of current methods, the analogues of the Hashin–Shtrikman bounds on κ^* and μ^* are now understood for fully anisotropic composites.[3,4,45] In particular, we now know those parts of the boundary of the G-closure which represent the 'strongest' and the 'weakest' anisotropic composites.

In the course of exploring these and other bounds for effective moduli, a number of powerful new tools have been introduced. The well-known Hashin–Shtrikman variational principles have been applied in new ways,[3,4,26,34,45] and new variational principles have been introduced, obtained from more classical ones by the addition of a quadratic null-Lagrangian.[5,27] In addition, entirely new approaches have been introduced: one is based on an equivalence between bounds for effective moduli and the lower semicontinuity of certain variational functionals;[30,32,63] another uses the fact that the effective moduli depend analytically on the component properties;[9,19,25,44] a third uses 'compensated compactness' to construct certain lower semicontinuous functionals,[17,18,35,37,66] and a fourth makes use of Hilbert space decompositions and continued fractions.[43] (These references represent a mere sampling of the relevant literature in each area.) The interested reader will find several of these new methods applied to a single problem in a self-contained manner in Ref. 27. The power and limitations of these various methods are just beginning to be understood, as are the relationships among them.[46]

To convey some of the flavor of these new developments, we present in detail one of the recently established bounds, an upper bound on the elastic energy quadratic form. There is of course a well-known bound due to Paul:[56]

$$(A^* \xi, \xi) \leqslant \theta_1 (A_1 \xi, \xi) + \theta_2 (A_2 \xi, \xi) \tag{3.2}$$

where θ_i is the volume fraction of the ith material, $i = 1, 2$. This bound is sharp, in the sense that for certain choices of the 'average strain' ξ there is a microstructure whose associated A^* achieves equality in (3.2). However, for *most* choices of ξ (3.2) is not saturated by any composite; therefore a better bound

$$(A^* \xi, \xi) \leqslant F(\theta_1, \theta_2, \mu_1, \mu_2, \kappa_1, \kappa_2, \xi) \tag{3.3}$$

is possible. We shall in fact prove the *optimal* bound of this type, in other words one which is saturated, for each ξ, by an appropriately chosen mixture of the two given materials. The method, which is based on the Hashin–Shtrikman variational principle, requires that the component materials be well ordered. Our presentation follows that of Ref. 26; equivalent results can be found presented somewhat differently in Refs. 3, 4 and 45. The function F on the right of (3.3) is given by (3.16) below, as the extremal value of a finite-dimensional mathematical programming problem.

As discussed in Section 2, it is sufficient to prove the bound for spatially periodic composites. We may therefore fix $Q = [0, 1]^d$ as the period cell; the microstructure is determined by the indicator functions $\chi_1(y)$ and $\chi_2(y) =$

$1 - \chi_1(y)$, $y \in Q$, constrained by the given volume fractions (2.12); and the effective Hooke's law is determined by (2.9).

The first step is to derive the Hashin–Shtrikman variational principle:

$$(A^* \xi, \xi) \leqslant -2 \int_Q (\sigma, \xi + e(\phi)) \chi_1 \, dy$$

$$+ \int_Q ((A_2 - A_1)^{-1} \sigma, \sigma) \chi_1 \, dy$$

$$+ \int_Q (A_2(\xi + e(\phi)), \, \xi + e(\phi)) dy \tag{3.4}$$

for any Q-periodic displacement field ϕ, and any Q-periodic field of symmetric tensors σ. The proof is elementary: expanding the pointwise inequality

$$|(A_2 - A_1)^{1/2} (\xi + e(\phi)) - (A_2 - A_1)^{-1/2} \sigma|^2 \geqslant 0 \tag{3.5}$$

and multiplying by χ_1 gives

$$-\chi_1((A_2 - A_1)(\xi + e(\phi)), \, \xi + e(\phi)) \leqslant$$
$$-2(\sigma, \xi + e(\phi)) \chi_1 + \chi_1((A_2 - A_1)^{-1} \sigma, \sigma) \tag{3.6}$$

The left side equals

$$((\tilde{A} - A_2)(\xi + e(\phi)), \, \xi + e(\phi)) \tag{3.7}$$

therefore integrating over Q and applying (2.9) we conclude (3.4).

The next step is to specialize to *constant* σ, and to evaluate the integrals in (3.4) wherever possible. This gives

$$((A^* - A_2)\xi, \xi) + 2\theta_1(\sigma, \xi) - \theta_1((A_2 - A_1)^{-1} \sigma, \sigma)$$

$$\leqslant -2 \int_Q (\sigma \chi_1, e(\phi)) dy + \int_Q (A_2 e(\phi), e(\phi)) dy \tag{3.8}$$

for any Q-periodic displacement field ϕ.

The third step is to minimize the expression on the right over ϕ. This amounts to solving the elastostatic equilibrium equation

$$\operatorname{div}(A_2 e(\phi)) - \operatorname{div}(\sigma \chi_1) = 0 \tag{3.9}$$

with a periodic boundary condition. It is convenient to use Fourier analysis: since A_2 and σ are constant, (3.9) determines the Fourier transform of ϕ at each frequency $k \in \mathbf{Z}^d$ directly in terms of the transform of χ_1 at the same frequency. After some algebra, one finds that the extremal value of the right side of (3.8) is

$$-\sum_{k \neq 0} |\hat{\chi}_1(k)|^2 \left(f\left(\frac{k}{|k|}\right)\sigma, \sigma \right) \tag{3.10}$$

where

$$\chi_1(y) = \sum_{k \in \mathbf{Z}^d} e^{2\pi i k \cdot y} \hat{\chi}_1(k) \tag{3.11}$$

and for any unit vector $v \in \mathbf{R}^d$, $f(v)$ is the 'degenerate Hooke's law' defined by

$$f(v)\sigma = \frac{d}{d\kappa_2 + 2(d-1)\mu_2}(\sigma v, v)v \odot v$$

$$+ \frac{1}{\mu_2}[(\sigma v) \odot v - (\sigma v, v)v \odot v] \tag{3.12}$$

Here σ is any symmetric tensor, and we use the notation $v \odot w = \frac{1}{2}(v \otimes w + w \otimes v)$ for the symmetric tensor product of two vectors in $\mathbf{R}^d$.

It remains to eliminate the explicit dependence of the bound on χ_1, which is after all arbitrary except for the volume fraction constraint. We use this constraint to see that

$$\int_Q (\chi_1 - \theta_1)^2 \, dy = \theta_1 \theta_2 \tag{3.13}$$

whence by Plancherel's theorem

$$\sum_{k \neq 0} |\hat{\chi}(k)|^2 = \theta_1 \theta_2 \tag{3.14}$$

This gives a bound on the 'nonlocal' term:

$$(3.10) \leqslant -\theta_1 \theta_2 \min_{|v|=1} (f(v)\sigma, \sigma) \tag{3.15}$$

Substitution into (3.8) gives a bound on A^* which still depends on the choice of a symmetric tensor σ, and minimization over σ gives a result of the desired form $(A^* \xi, \xi) \leqslant F$, with

$$F = (A_2 \xi, \xi) + \theta_1 \cdot \min_\sigma \{ -2(\sigma, \xi) + ((A_2 - A_1)^{-1}\sigma, \sigma)$$

$$- \theta_2 \min_{|v|=1} (f(v)\sigma, \sigma) \} \tag{3.16}$$

Our interest in this bound lies in the fact that it is the *best possible* bound for $(A^* \xi, \xi)$ in terms of the given parameters $\xi, \theta_1, \theta_2 = 1 - \theta_1$, and the bulk and shear moduli of the component materials $\kappa_1 \leqslant \kappa_2, \mu_1 \leqslant \mu_2$. This will be proved in the next section, as an application of the formula for the effective behavior of a sequentially laminated composite.

4. CONSTRUCTION OF MIXTURES WITH EXPLICITLY COMPUTABLE EFFECTIVE MODULI

For most microstructures there is no explicit, algebraic formula for the tensor of effective moduli A^*; one must make do instead with a variational principle such as (2.9) or (2.11), or perhaps with the partial differential equation characterizing its extremal. If this were the only available tool it would be virtually impossible to establish the optimality of any bound! Fortunately there are certain, rather special microstructures for which the effective moduli *are* computable; and, remarkably, this class of composites is rich enough to demonstrate the optimality of a variety of bounds, including (3.3).

There are some simple and more or less classical examples of composites with explicitly computable properties. One example is that of a layered microstructure;[6,11,40] another is the 'concentric sphere construction', which was used by Hashin in Ref. 73 to prove the optimality of their bulk modulus bounds. It is natural enough to iterate such constuctions, for example layering together two composites each of which has its own fine-scale structure, obtained perhaps by layering or by a version of the concentric sphere construction. This idea, which can be found in Bruggeman's work,[11] has been rediscovered by various individuals and applied to prove the attainability of many different bounds, e.g. Refs. 3–5, 17, 18, 26, 34, 35, 37, 38, 42, 61, 66.

An important new development concerns the attainability of certain mean field theories. The formulas they predict for the tensor of effective moduli A^* were originally intended as approximations, not as exact results. Nevertheless, it has recently been shown that certain effective medium theories are *exactly* attainable by composites with appropriately chosen microstructures.[2,36,41] Obviously, this result greatly expands the class of composites with explicitly computable effective moduli—particularly since these effective medium theories (the 'coherent potential approximation' and the 'differential effective medium theory') have been widely studied in the mechanics literature, see e.g. Refs. 74, 75.

The microstructures that arise from these constructions are, it should be understood, somewhat idealized materials. They are higly ordered, neither periodic nor stochastic in character, and they frequently involve multiple length scales. It may seem like cheating that we allow the use of such microstructures to establish the attainability of a bound, whereas the proof of the bound may make use of a special structure such as periodicity. This is in fact perfectly legitimate; indeed, it is here that we use the power of the mathematical theory. The point is that these constructions fit perfectly into the mathematical context of G-convergence (see especially Ref. 2); therefore, by the 'density' result mentioned at the end of Section 2, their effective moduli can be approximated arbitrarily well by those of spatially periodic composites. Actually, it is quite natural to use the most restrictive possible setting for proving bounds, and the most general one for showing that they are achieved.

The remainder of this section is devoted to a discussion of *sequentially laminated* composites, and to a proof of the attainability of the new upper bound (3.3). Closely related ideas and results can be found in Refs. 3, 4, 26, 45. The construction of a sequentially laminated composite is an iterative procedure, producing a microstructure that has several different length scales. A *laminar composite of rank 1* is obtained by layering two initially given materials, specifying the proportion of each and the layer direction, and using a small parameter ε_1 as the layer thickness. As $\varepsilon_1 \to 0$, the elastic behavior is described by an effective Hooke's law C_1. A *laminar composite of rank 2* is obtained by layering two laminar composites of rank 1, again specifying the proportion of each and the layer direction, and using another small parameter ε_2 for the layer thickness. As $\varepsilon_1, \varepsilon_2 \to 0$ with $\varepsilon_1 \ll \varepsilon_2$, the elastic behavior is described by an effective Hooke's law C_2. This process can clearly be continued indefinitely: the general sequentially laminated composite of rank r is obtained by layering together two sequentially laminated composites of rank $r-1$. We shall consider here only a special case, in which *one of these two materials is the isotropic one with shear modulus μ_2 and bulk modulus κ_2 at each successive stage.* An elegant, iterative formula for representing the effective moduli of such a composite was derived in Ref. 17, following a method developed for scalar equations in Ref. 66. We now give a derivation of this result.

The basic building block is a formula for the effective tensor C corresponding to a layered mixture of the isotropic material with Hooke's law A_2 and a general elastic material with Hooke's law B, using layers orthogonal to the unit vector $v \in \mathbf{R}^n$, and using proportions ρ_2 and $\rho_B = 1 - \rho_2$ of A_2 and B, respectively:

$$\rho_B (A_2 - C)^{-1} \sigma = (A_2 - B)^{-1} \sigma - \rho_2 f(v) \sigma \qquad (4.1)$$

for any symmetric tensor σ. Here $f(v)$ is the *same* degenerate Hooke's law that arose in our proof of the bound, defined by (3.12). In writing (4.1) we have implicitly assumed that $A_2 - C$ and $A_2 - B$ are invertible, when viewed as symmetric linear maps on the space of symmetric tensors. This is the case whenever $B < A_2$, since then $C < A_2$ as well, by Paul's bound (3.2); this ordering hypothesis will be sufficient for our purposes, since we are concerned with mixtures of two well-ordered isotropic materials, i.e. (3.1) holds. (There is a version of (4.1) without invertibility hypotheses, see for example Ref. 17.) To prove (4.1), one must of course begin with a characterization of C. In a layered composite of the type under consideration, the local values of the stress and strain are essentially constant within each component. Therefore, arguing for example as in Ref. 40, the calculation of $C\xi$ given ξ is easily reduced to this algebraic problem: find a pair of symmetric matrices ξ_2 and ξ_B (representing the strain in the layers occupied by materials A_2 and B respectively) such that

$$\rho_2 \xi_2 + \rho_B \xi_B = \xi$$

$$\xi_B - \xi_2 = v \odot w \text{ for some } w \in \mathbf{R}^n \qquad (4.2\text{a--c})$$

$$(A_2 \xi_2 - B\xi_B)v = 0$$

The first relation says that ξ is the average strain; the second is the consistency condition for the existence of a deformation with the specified piecewise constant strain (recall that $v \odot w = (v \otimes w + w \otimes v)/2$); and the third represents the continuity of the normal stress at the layer interface. In terms of these quantities, $C\xi$ is determined by

$$C\xi = \rho_2 A_2 \xi_2 + \rho_B B\xi_B \qquad (4.2\text{d})$$

which identifies it as the average stress. The solution of (4.2a–d) is easiest to represent in terms of $\sigma = (A_2 - C)\xi$. One calculates that ξ_2 and ξ_B are given in terms of σ by

$$\xi_B = \rho_B^{-1}(A_2 - B)^{-1}\sigma, \quad \xi_2 = \xi_B - v \odot w \qquad (4.3)$$

where $w \in \mathbf{R}^n$ is chosen so that

$$\rho_B A_2(v \odot w) = 2(\sigma v) \odot v - (\sigma v, v)v \odot v \qquad (4.4)$$

whence

$$\rho_B[A_2(v \odot w)]v = \sigma v \qquad (4.5)$$

The unique w satisfying (4.4) is

$$w = \rho_B^{-1}\left[\frac{d}{d\kappa_2 + 2(d-1)\mu_2}(\sigma v, v)v + \frac{1}{\mu_2}(\sigma v - (\sigma v, v)v)\right] \qquad (4.6)$$

and it has the property that

$$\rho_B v \odot w = f(v)\sigma \tag{4.7}$$

with $f(v)$ defined by (3.12). Therefore

$$\begin{aligned}
(A_2 - C)^{-1}\sigma = \xi &= \rho_B \xi_B + \rho_2 \xi_2 \\
&= \xi_B - \rho_2 v \odot w \\
&= \rho_B^{-1}(A_2 - B)^{-1}\sigma - \rho_B^{-1}\rho_2 f(v)\sigma
\end{aligned} \tag{4.8}$$

which is precisely the desired formula (4.1).

Now consider a sequence $C_0, C_1, C_2, \ldots$ of effective tensors such that

$C_0 = A_1$ represents an isotropic material with bulk
modulus κ_1 and shear modulus μ_1, $\qquad$ (4.9a)

and, for $r \geq 1$,

C_r is obtained by layering A_2 with C_{r-1} in volume
fractions α_r and $(1 - \alpha_r)$ respectively, using the unit
vector v_r as the layer normal. $\qquad$ (4.9b)

Evidently, C_r represents the effective behavior of a certain sequentially laminated composite of rank r. The volume fraction of A_2 in C_r is

$$\beta_r = 1 - \prod_{i=1}^{r}(1 - \alpha_i), \quad r \geq 1; \quad \beta_0 = 0 \tag{4.10}$$

A formula for C_r is easily obtained by iterating (4.1):

$$(1 - \beta_r)(A_2 - C_r)^{-1} = (A_2 - A_1)^{-1} - \sum_{i=1}^{r}(\beta_i - \beta_{i-1})f(v_i) \tag{4.11}$$

Let us terminate this process at $r = N$, and write

$\theta_2 = \beta_N = $ overall volume fraction of A_2

$\theta_1 = 1 - \beta_N = $ overall volume fraction of A_1 $\qquad$ (4.12)

$A^* = C_N = $ effective Hooke's law of the associated rank N
composite.

It is easy to see that the sequence

$$m_i = (\beta_i - \beta_{i-1})/\beta_N, \quad 1 \leq i \leq N \tag{4.13}$$

can be any nonnegative sequence which sums to 1, by making an appropriate choice of the parameters $\{\alpha_i\}$. Thus we have shown that *for any integer $N \geq 1$, any unit vectors $\{v_i\}_{i=1}^{N}$ in $\mathbf{R}^d$, any real numbers $\{m_i\}_{i=1}$ with $0 \leq m_i \leq 1$ and $\Sigma m_i = 1$ and any real number $\theta_2, 0 < \theta_2 < 1$, there is a sequen-*

tially laminated composite made by mixing A_1 and A_2 as in (4.9), using overall volume fractions $\theta_1 = 1 - \theta_2$ and θ_2 respectively, whose effective Hooke's law A^* is characterized by

$$\theta_1(A_2 - A^*)^{-1} = (A_2 - A_1)^{-1} - \theta_2 \sum_{i=1}^{N} m_i f(v_i) \qquad (4.14)$$

We now apply this construction to establish the optimality of the new upper bound (3.16) Our task is to show that for each symmetric tensor ξ there is a choice of the parameters $\{v_i, m_i\}$ such that A^*, defined by (4.14), satisfies $(A^* \xi, \xi) = F$ with F as in (3.16). Now, (3.16) gives F in terms of a mathematical programming problem

$$\min_{\sigma} \{ -2(\sigma, \xi) + ((A_2 - A_1)^{-1} \sigma, \sigma) - \theta_2 \min_{|v| = 1} (f(v)\sigma, \sigma) \} \qquad (4.15)$$

over symmetric tensors σ, so it is reasonable to expect the proper choices of $\{v_i, m_i\}$ to emerge from the optimality conditions for (4.15). Since the last term is not a smooth function of σ, it is natural to use the methods of 'nonsmooth analysis', see for example Ref. 15. To this end we rewrite (4.15) as

$$\min_{\sigma} \{ -2(\sigma, \xi) + g(\sigma) \} \qquad (4.16)$$

with

$$g(\sigma) = \max_{|v| = 1} ((A_2 - A_1)^{-1} - \theta_2 f(v)\sigma, \sigma) \qquad (4.17)$$

For each fixed v the expression on the right is a positive, quadratic function of σ (one way to establish positivity is to make use of (4.1)). Therefore g is convex, and the optimality condition for (4.17) is that for any extremal σ^*

$$2\xi \in \partial g(\sigma^*) \qquad (4.18)$$

where $\partial g(\sigma^*)$ is the subdifferential of g at σ^* (see for example Ref. 15, 2.3.1–2.3.3 and Corollary 1, §2.3). Moreover, $\partial g(\sigma^*)$ is the closed convex hull of the differentials of the various quadratic forms in (4.17) as v ranges over all extremals (see for example, Ref. 15, §2.8, Corollary 1). Since the space of symmetric tensors is finite dimensional, each element of the closed convex hull is in fact a convex combination of finitely many extreme points. Therefore the optimality condition (4.18) becomes

$$\xi = (A_2 - A_1)^{-1} \sigma^* - \theta_2 \sum_{i=1}^{N} m_i f(v_i) \sigma^* \qquad (4.19)$$

with $m_i \geqslant 0$, $\Sigma m_i = 1$, $|v_i| = 1$, $N < \infty$, and

$$g(\sigma^*) = ((A_2 - A_1)^{-1} \sigma^*, \sigma^*) - \theta_2(f(v_i)\sigma^*, \sigma^*), \quad 1 \leqslant i \leqslant N \qquad (4.20)$$

Comparing (4.19) with (4.14), we see that

$$\xi = \theta_1(A_2 - A^*)^{-1} \sigma^* \qquad (4.21)$$

where A^* is the sequentially laminated composite of rank N constructed using $\{m_i, v_i\}_{i=1}^N$. We claim that this A^* satisfies $(A^* \xi, \xi) = F$. Indeed, the value of F is

$$F = (A_2 \xi, \xi) + \theta_1 \{-2(\sigma^*, \xi) + g(\sigma^*)\} \qquad (4.22)$$

using (3.16) and the fact that σ^* is extremal for (4.15). We have

$$(\sigma^*, \xi) = g(\sigma^*) \qquad (4.23)$$

by (4.19) and (4.20), so (4.22) becomes

$$F = (A_2 \xi, \xi) - \theta_1(\sigma^*, \xi) \qquad (4.24)$$

But $\theta_1 \sigma^* = (A_2 - A^*)\xi$ by (4.21), and substitution gives the desired result $F = (A^* \xi, \xi)$.

5. APPLICATIONS TO STRUCTURAL OPTIMIZATION

The recent interest in optimal bounds on the effective moduli of composites has been stimulated in large part by applications to structural optimization, see for example, Refs. 1, 28, 38, 39, 49, 50, 67. That discipline is concerned with choosing the geometry or composition of a load-bearing structure so as to use the available materials as efficiently as possible. The subject has a rich history and an extensive literature; books and articles reviewing various aspects include Refs. 7, 21, 53, 57. Initially attention was focused primarily on analytical methods—optimality conditions, conformal mapping, isoperimetric inequalities, and so forth. More recently, with the growing feasibility of large scale computing, attention has naturally been turned to methods for the direct, numerical calculation of optimal structures.

To fix ideas, let us consider a particular problem involving shape optimization and plane stress. We begin with a homogeneous, isotropic elastic body occupying a region $\Omega \subset \mathbf{R}^2$, loaded along its boundary $\partial \Omega$ by a specified traction f. We desire to lighten this body by removing material from a subset $H \subset \Omega$, consisting of one or more holes of arbitrary size and shape. The goal is to achieve the *minimum possible weight*, i.e. to maximize the area of the 'holes' H, subject to a *performance constraint* on the stress σ_H or displacement u_H of the resulting elastic structure. Typical constraints are

that the work done by the load ('compliance')

be not too large: $\displaystyle\int_{\partial\Omega} u_H \cdot f \leqslant C;$ or $\hspace{3cm}$ (5.1a)

that the average displacement on a subdomain

Ω_1 be not too large: $\displaystyle\int_{\Omega_1} |u_H| \leqslant C;$ or $\hspace{2.5cm}$ (5.1b)

that the pointwise maximum stress be not too

large: $\displaystyle\sup_{x\in\Omega} \| \sigma_H(x) \| \leqslant C$ $\hspace{4.5cm}$ (5.1c)

Highly efficient and sophisticated algorithms have been developed for the numerical solution of such problems; Ref. 21 gives an excellent review. Typically, one begins by deciding how many holes to consider. Each hole boundary is determined by finitely many points, for example using splines. The resulting domain is triangulated, and the equations of elastostatics are modeled as a finite system of linear equations using the finite element method. The design problem is thus transformed to a (highly nonlinear!) mathematical programming problem, and one can seek an 'optimal' design—or at least an improvement of a given design—using steepest descent, or perhaps some more sophisticated method.

Though its utility is beyond dispute, this 'conventional' approach has one troublesome aspect: the gross features of the design—especially, the number of holes—must be chosen at the outset; they are not a part of the optimization. Thus the output is likely to be a local optimum, or at best an optimum among all designs with a specified number of holes. In fact, numerical attempts at *global* optimization for related model problems have led in some cases to 'optimal' designs that vary on the scale of the mesh size itself, with no convergence evident as the mesh size tends to zero![1,13] This phenomenon is now well understood. In the context of shape optimization, the situation is as follows: consider first the best design with one hole, then that with two, and so forth. As the number of holes gets larger the performance may get better (depending of course, on the specific problem under consideration). In the limit of infinitely many holes one thus finds a global optimum which is not a 'conventional' design at all, but instead a structure made from *composite materials obtained by perforation.*

With hindsight it seems almost obvious: if one is prepared to consider designs with many small holes, then one ought also to consider their limits. We thus arrive at a new approach to structural optimization: if the goal is to find a global optimum then it is best to work from the start in the class of all *structures made up of composite materials obtainable by perforation from the one initially given*. It should be emphasized that the underlying problem is *not* being changed, since we allow *only* composites achievable by perforation, and we are careful to model them properly. However, the resulting optimization problem looks quite different: whereas initially we were considering structures made up of a single material (or the absence thereof), now we propose to allow a continuum of materials—each representing a perforated composite with a different microscopic geometry. (As a technical matter, the mathematical theory discussed in the preceding sections does not quite apply to perforated composites, since it requires $\mu_i > 0$ and $\kappa_i > 0$. This can be circumvented, at least for compliance optimization problems, by the methods of Refs. 30, 32. Alternatively, we can simply treat the 'holes' as though they were filled with a very weak elastic material.)

The introduction of composites as generalized designs—sometimes called the *relaxation* of the design problem—has been studied extensively by several groups over the past ten years, see for example, Refs. 20, 30, 38, 50, 52, 54, 58, 59, 67. From a theoretical standpoint, the principal advantage of relaxation is that it assures the *existence* of an optimal design; roughly, this means that a numerical solution of the relaxed problem will converge as the mesh size tends to zero. There is also a practical advantage, based on the fact that the initial material and the absence of material are included (as extreme cases) among the candidate composites: evidently, for a given finite element subdivision the introduction of composites serves to enlarge the design space and hence to *improve the performance* of a numerically obtained optimal design. Moreover, precisely because it has the effect (within a finite element context) of enlarging the design space, the process of relaxation can *destroy local minima*—making it easier to locate a globally optimal design. Finally, since the relaxed problem is known to have a solution, it is meaningful to use the associated *optimality conditions*; this has led in some contexts to closed-form examples of optimal designs making use of composites, for example, Refs. 29–31. The method of relaxation has its limitations: the optimal designs obtained this way may be difficult or even impossible to manufacture, because of the presence of composites. Even so, these solutions can be used as *benchmarks* against which to compare the output of a more conventional algorithm.

The process of relaxation is conceptually simple: we must simply reformulate the design problem in a form that permits perforated composites as admissible materials. The actual execution, however, is not so simple: it requires specific knowledge about the properties of the relevant composites. For a local performance criterion such as the maximum stress (5.1c) we would have to know optimal bounds relating the effective Hooke's law, the density of holes, the average stress, and the local maximum stress in a general perforated composite. This represents a challenge for the future: no such result is presently known. For a performance criterion involving some integral of the displacement, such as (5.16), it would suffice to know the solution of the G-closure problem—in other words, to know the class of all effective Hooke's laws obtainable using perforations that remove a given fraction of the material. The analogous problem has been solved for scalar equations,[37,66] and it has been applied to solve various optimization problems involving conductivity, see for example, Refs. 12, 20, 30, 38, 50, 67; but unfortunately the G-closure problem for elasticity remains open at this time except in certain rather special cases.[34,35] However, problems involving compliance constraints such as (5.1a) do not require the full solution of the G-closure problem; rather, bounds of the type presented in Sections 3 and 4 are sufficient. To explain why, we note that it is not really necessary to consider *all* composites; one might as well consider just those that can actually occur in an optimal design. Now, by Green's formula the compliance is equal to the internal elastic energy:

$$(5.2) \qquad \int_{\partial\Omega} u \cdot f = \int_{\Omega} (A(x)e(u), e(u)) \, dx$$

where $A(x)$ is the spatially varying tensor of elastic moduli and u the associated displacement. A structure which minimizes weight for fixed compliance will also minimize compliance for given weight; it is not hard to see from this that $A(x)$ should *maximize* $(Ae(u), e(u))$ at each point x in an optimal design. Thus the values that $A(x)$ can take in an optimal design are restricted to those that maximize $(A\xi, \xi)$ for some tensor ξ.

The preceding discussion shows that we have enough information to solve optimal design problems with compliance constraints, but it falls short of specifying an algorithm to do so. How, operationally, should one proceed? Following Ref. 30, we advocate an algorithm based on the principle of minimum complementary energy, a variational principle for the stress whose extremal value is equal to the compliance:

$$\int_{\partial\Omega} u \cdot f = \min_{\text{div } \sigma = 0, \, \sigma \cdot n = f} \int_{\Omega} (A^{-1}(x)\sigma, \sigma) \, dx \qquad (5.3)$$

Introducing a Lagrange multiplier for the performance constraint (5.1a) our design problem is

$$\text{MIN}_{\text{designs}}\{\text{WEIGHT} + \lambda \cdot \text{COMPLIANCE}\} \tag{5.4}$$

The outer minimization over designs is quantified by introducing functions $\theta(x)$ and $A(x)$, the density and effective Hooke's law, constrained by the pointwise conditions

$0 \leqslant \theta \leqslant 1$, and A is the effective Hooke's law of a
perforated composite obtained by removing volume
fraction $1 - \theta$ of the initially given material.   (5.5)

The compliance is itself a minimum, according to (5.3), so (5.4) becomes

$$\min_{\theta, A} \left\{ \int_\Omega \theta(x)\,dx + \lambda \cdot \min_{\text{div }\sigma = 0, \sigma \cdot n = f} \int_\Omega (A^{-1}(x)\sigma, \sigma)\,dx \right\} \tag{5.6}$$

The order of minimization is unimportant, and switching it gives

$$\min_{\text{div }\sigma = 0, \sigma \cdot n = f} \int_\Omega \Phi_\lambda(\sigma)\,dx \tag{5.7}$$

with

$$\Phi_\lambda(\sigma) = \min_{\theta, A} [\theta + \lambda(A^{-1}\sigma, \sigma)] \tag{5.8}$$

The minimization in (5.8) is over real numbers θ and tensors A, constrained by (5.5). This is slightly different from the problem we treated in Sections 3 and 4, but *it can be solved by exactly the same method*—as can considerably more general problems, for example the analogue of (5.8) when there are compliance constraints under two or more loads.

The next step, of course, is to evaluate (5.8) analytically or numerically, and to carry out the optimization by solving (5.7) for realistic design problems. Work in these directions is currently in progress. The minimization of (5.8) was executed in Ref. 30 for the special case of an elastic material in plane stress with Poisson's ratio zero, i.e. when $\mu = \kappa = \frac{1}{2}E$, where E is Young's modulus, using a different method, based on quasiconvexification. The answer is surprisingly simple: scaling $\lambda = E = 1$ for simplicity,

$$\Phi_1(\sigma) = \begin{cases} 1 + \sigma_1^2 + \sigma_2^2, & |\sigma_1| + |\sigma_2| \geqslant 1 \\ 2(|\sigma_1| + |\sigma_2|) - 2|\sigma_1 \sigma_2|, & |\sigma_1| + |\sigma_2| \leqslant 1 \end{cases}$$

where σ_1 and σ_2 are the principal stresses (the eigenvalues of σ).

ACKNOWLEDGEMENTS

This work was supported in part by NSF grant DMS-8312229, ONR grant N00014-83-K-0536, DARPA contract F49620-87-C-0065, and the Sloan Foundation.

REFERENCES

1. ARMAND, J.-L., LURIE, K. A. and CHERKAEV, A. V., Optimal control theory and structural design, in: *New Directions in Optimum Structural Design* (E. Atrek *et al.*, eds), New York, John Wiley, 1984, p. 211.
2. AVELLANEDA, M., Iterated homogenization, differential effective medium theory, and applications, *Comm. Pure Appl. Math.*, **40** (1987), 527.
3. AVELLANEDA, M., Optimal bounds and microgeometries for elastic composites, *SIAM J. Appl. Math.*, **47** (1987), 1216.
4. AVELLANEDA, M., Bounds on the effective elastic constants of two-phase composite materials, to appear in *Proc. Sem. Collège de France*.
5. AVELLANEDA, M., CHERKAEV, A. V., LURIE, K. A. and MILTON, G. W., On the effective conductivity of polycrystals and a three-dimensional phase inter-change inequality, *J. Appl. Phys.*, (1988) (in press).
6. BACKUS, G. E., Long-wave elastic anisotropy produced by horizontal layering, *J. Geophys. Res.*, **67** (1962), 4427.
7. BANICHUK, N. V., *Problems and Methods of Optimal Structural Design*, New York, Plenum Press. 1983.
8. BENSOUSSAN, A., LIONS, J.-L. and PAPANICOLAOU, G., *Asymptotic Analysis for Periodic Structures*, Amsterdam, North-Holland, 1978.
9. BERGMAN, D. J., The dielectric constant of a composite material—a problem in classical physics, *Phys. Rep.*, **C43** (1978), 377.
10. BERRYMAN, J. G. and MILTON, G. W., Microgeometries of random composites and porous media, *J. Phys. D* (1988) (in press).
11. BRUGGEMAN, D. A. G., Berechnung verschiedener physikalischer konstanten, von heterogenen substanzen, *Ann. Phys.*, **5** (1935) 636; also Elastizität kons-tanten von kristallaggragaten, Ph.D. Thesis, Utrecht, 1930.
12. CABIB, E. and DAL MASO, G., On a class of optimum problems in structural design, *J. Opt. Th. Appl.*, **56** (1988) (in press).
13. CHENG, K.-T. and OLHOFF, N., An investigation concerning optimal design of solid elastic plates, *Int. J. Solids Struct.*, **17** (1981), 305.
14. CHRISTENSEN, R. M., *Mechanics of Composite Materials*, New York, Wiley Interscience, 1979.
15. CLARKE, F. H., *Optimization and Nonsmooth Analysis*, New York, John Wiley, 1983.
16. DAL MASO, G. and KOHN, R., The local character of G-closure, in preparation.
17. FRANCFORT, G. A. and MURAT, F., Homogenization and optimal bounds in linear elasticity, *Arch. Rat. Mech. Anal.*, **94** (1986), 307.

18. GIBIANSKI, L. V. and CHERKAEV, A. V., Design of composite plates of extremal rigidity, Ioffe Physicotechnical Institute preprint, 1984.
19. GOLDEN, K. and PAPANICOLAOU, G., Bounds for effective parameters of heterogeneous media by analytic continuation, *Comm. Math. Phys.*, **90** (1983), 473.
20. GOODMAN, J., KOHN, R. V. and REYNA, L., Numerical study of a relaxed variational problem from optimal design, *Comp. Meth. Appl. Mech. Engng*, **57** (1986), 107.
21. HAFTKA, R. T. and GRANDHI, R. V., Structural shape optimization—a survey, *Comp. Meth. Appl. Mech. Engng*, **57** (1986), 91.
22. HASHIN, Z., Analysis of composite materials: a survey, *J. Appl. Mech.*, **50** (1983), 481.
23. HASHIN, Z. and SHTRIKMAN, S., A variational approach to the theory of the elastic behavior of multiphase materials, *J. Mech. Phys. Solids*, **11** (1963), 127.
24. HILL, R., Elastic properties of reinforced solids: some theoretical principles, *J. Mech. Phys. Solids*, **11** (1963), 357.
25. KANTOR, Y. and BERGMAN, D. J., Improved rigorous bounds on the effective elastic moduli of a composite material, *J. Mech. Phys. Solids*, **32** (1984), 41.
26. KOHN, R. V. and LIPTON, R., Optimal bounds for the effective energy of a mixture of two incompressible elastic materials, *Arch. Rat. Mech. Anal.* (1988) (in press).
27. KOHN, R. V. and MILTON, G. W., On bounding the effective conductivity of anisotropic composites, in: *Homogenization and Effective Moduli of Materials and Media*, J. Ericksen *et al.*, eds., Berlin, Springer-Verlag, 1986, p. 97.
28. KOHN, R. V. and STRANG, G., Structural design optimization, homogenization, and relaxation of variational problems, in: *Macroscopic Properties of Disordered Media,* (R. Burridge *et al.*, eds), Berlin, Springer-Verlag, 1982, p. 131.
29. KOHN, R. and STRANG, G., Optimal design for torsional rigidity, in: *Hybrid and Mixed Finite Element Methods* (S. N. Atluri *et al.* eds), New York, John Wiley, 1983, p. 281.
30. KOHN, R. and STRANG, G., Optimal design and relaxation of variational problems I–III, *Comm. Pure Appl. Math.,* **39**, (1986), 113, 139, 353.
31. KOHN, R. V. and STRANG, G., The constrained least gradient problem, in: *Non-Classical Continuum Mechanics* (R. J. Knops and A. A. Lacey, eds), Cambridge, Cambridge University Press, 1987, p. 226.
32. KOHN, R. V. and VOGELIUS, M., Relaxation of a variational method for impedance computed tomography, *Comm. Pure Appl. Math.*, **40** (1987), 745.
33. KOZLOV, S. M., The averaging of random operators, *Math. USSR – Sbornik*, **37** (1980), 167.
34. LIPTON, R., On the effective elasticity of a two dimensional homogenized incompressible elastic composite, *Proc. Roy. Soc. Edinburgh*, Ser. A, (1988) (in press).
35. LURIE, K. A. and CHERKAEV, A. V., G-closure of some particular sets of admissible material characteristics for the problem of bending of thin elastic plates, *J. Opt. Th. Appl.* **42** (1984), 305.
36. LURIE, K. A. and CHERKAEV, A. V., Optimization of properties of multi-component isotropic composites, *J. Opt. Th. Appl.*, **46** (1985), 571; also The

problem of formation of an optimal isotropic multicomponent composite, Ioffe Physicotechnical Institute preprint No. 895.

37. LURIE, K. A. and CHERKAEV, A. V., Exact estimates of the conductivity of a binary mixture of isotropic components, *Proc. Roy. Soc. Edinburgh*, Ser. A., **104** (1986), 21.

38. LURIE, K. A. and CHERKAEV, A. V., The effective properties of composites and problems of optimal design of constructions (in Russian), *Uspekhi Mekhaniki*, No. 2, 1987.

39. LURIE, K. A., CHERKAEV, A. V. and FEDOROV, A. V., Regularization of optimal design problems for bars and plates I, II, *J. Opt. Th. Appl.*, **37** (1982), 499, 523.

40. MCCONNELL, W. H., On the approximation of elliptic operators with discontinuous coefficients, *Ann. Sci. Norm. Sup. Pisa*, **3** (1976), 121.

41. MILTON, G. W., The coherent potential approximation is a realizable effective medium theory, *Comm. Math. Phys.*, **99** (1985), 465.

42. MILTON, G. W., Modeling the properties of composites by laminates, in: *Homogenization and Effective Moduli of Materials and Media* (J. Ericksen et al., eds), Berlin, Springer-Verlag, 1986, p. 150.

43: MILTON, G. W., Multicomponent composites, electrical networks, and new types of continued fractions I, II, *Comm. Math. Phys.* (1988) (in press).

44. MILTON, G. W. and GOLDEN, K., Thermal conduction in composites, in: *Thermal Conductivity 18* (T. Ashworth and D. R. Smith, eds), New York, Plenum Press, 1985, p. 571.

45. MILTON, G. W. and KOHN, R. V., Variational bounds on the effective moduli of anisotropic composites, *J. Mech. Phys. Solids* (in press).

46. MILTON, G. W. and MCPHEDRAN, R. C., A comparison of two methods for deriving bounds on the effective conductivity of composites, in: *Macroscopic Properties of Disordered Media* (R. Burridge et al., eds), Berlin, Springer-Verlag, 1982, p. 183.

47. MILTON, G. W. and PHAN-THIEN, N., New bounds on the effective elastic moduli of two-dimensional materials, *Proc. Roy. Soc. London*, **A380** (1982), 305.

48. MURAT, F., H-convergence, mimeographed lecture notes, Université d'Alger, 1978.

49. MURAT, F., Control in coefficients, in: *Systems and Control Theory Encyclopedia: Theory, Technology, Applications*, Oxford, Pergamon Press, 1986.

50. MURAT, F. and TARTAR, L., Calcul des variations et homogénéization, in: *Les Méthodes de l'Homogenéization: Theorie et Applications en Physique*, Coll. de la Dir. des Etudes et Recherches d'Electricité de France, Eyrolles, 1985, p. 319.

51. NORRIS, A. N., A differential scheme for the effective moduli of composites, *Mech. Mater.*, **4** (1985), 1.

52. OLHOFF, N., LURIE, K. A., CHERKAEV, A. V. and FEDOROV, A. V., Sliding regimes and anisotropy in optimal design of vibrating axisymmetric plates, *Int. J. Solids Struct.*, **17** (1981), 931.

53. OLHOFF, N. and TAYLOR, J., On structural optimization, *J. Appl. Mech.*, **50**, (1983), 1139.

54. ONG, T.-G., ROZVANY, G. I. N. and SZETO, W. T., Least weight design of perforated elastic plates for given compliance: non-zero Poisson's ratio, to appear.

55. PAPANICOLAOU, G. and VARADHAN, S. R. S., Boundary value problems with rapidly oscillating random coefficients, in: *Colloquia Mathematica Societatis*

Janos Bolyai, Vol. **27**, Random Fields, Amsterdam, North Holland, 1982, p. 835.

56. PAUL, B., Prediction of elastic constants of multiphase materials, *Trans. ASME*, **218** (1960), 36.

57. PIRONNEAU, O., *Optimal Shape Design for Elliptic Systems*, Berlin, Springer-Verlag, 1984.

58. RAITUM, U. E., On optimal control problems for linear elliptic equations, *Soviet Math. Dokl.*, **20** (1979), 129.

59. ROZVANY, G. I. N., ONG, T. G., SZETO, W. T., SANDLER, R., OLHOFF, N. and BENDSOE, M. P., Least-weight design of perforated elastic plates I, II, *Int. J. Solids Struct.*, **23** (1987), 521, 537.

60. SANCHEZ-PALENCIA, E., *Non-homogeneous Media and Vibration Theory*, Lecture Notes in Physics 127, Berlin, Springer-Verlag, 1980.

61. SCHULGASSER, K., Relationship between single-crystal and polycrystal electrical conductivity, *J. Appl. Phys.*, **47** (1976), 1880.

62. SPAGNOLO, S., Convergence in energy for elliptic operators, in: *Numerical Solution of Partial Differential Equations III Synspade 1975* (B. Hubbard, ed.), New York, Academic Press, 1976.

63. STRANG, G. and KOHN, R. V., Optimal design of a two-way conductor, in: *Non-Smooth Mechanics* (P. D. Panagiotopoulos *et al.*, eds), Birkhaeuser Verlag, Basel, 1988.

64. SUQUET, P., Une méthode duale en homogénéisation: application aux milieux élastiques, *J. Mech. Theor. Appl.*, special issue (1982), 79.

65. TARTAR, L., Cours Peccot, Collège de France, 1977.

66. TARTAR, L., Estimations fines des coefficients homogénéisés, in: *Ennio de Giorgi's Colloquium* (P. Krée ed.), London, Pitman, 1985, p. 168.

67. TARTAR, L., The appearance of oscillations in optimization problems, in: *Non-Classical Continuum Mechanics* (R. J. Knops and A. A. Lacey, eds), Cambridge, Cambridge University Press, 1987, p. 129.

68. WILLIS, J. R., Variational and related methods for the overall properties of composite materials, in: *Advances in Applied Mechanics*, Vol. 21, (C. S. Yih, ed.), 1981, p. 2.

69. WILLIS, J. R., Elasticity theory of composites, in: *Mechanics of Solids* (H. G. Hopkins and M. J. Sewell, eds), Oxford, Pergamon Press, 1982, p. 653.

70. YURINSKII, V. V., Average of an elliptic boundary value problem with random coefficients, *Siberian Math. J.*, **21** (1980), 470.

71. ZHIKOV, V. V., KOZLOV, S. M., OLEINIK, O. A. and NGOAN, K. T., Averaging and G-convergence of differential operators, *Russian Math. Surveys*, **34** (1979), 69.

72. WATT, P. J., The elastic properties of composite materials, *Rev. Geophys. and Space Phys.*, **14** (1976), 541.

73. HASHIN, Z., The elastic moduli of heterogeneous materials, *ASME J. Appl. Mech.*, **29** (1962), 143.

74. BERRYMAN, J. G., Long wavelength propagation in composite elastic media I, II, *J. Acoust. Soc. Amer.*, **68** (1980), 1809.

75. ELLIOTT, R. J., KRUMHANSL, J. A. and LEATH, P. L., The theory and properties of randomly disordered crystals and related physical systems, *Rev. Mod. Phys.*, **46** (1974), 465.

List of Participants

ARCISZ, M.
IPPT Pan,
Swietokrzysha 21,
00-499 Warsaw, Poland

BAO, G.
Center for the Application
of Mathematics,
Lehigh University,
Bethlehem, Pennsylvania 18015,
USA

BARSOUM, R. S.
US Army Laboratory Command,
Materials Technology
Laboratory,
Watertown, Massachusetts
02172–0001, USA

BOCCA, P.
Istituto Universitario di
Architettura di Venezia
30125 Venezia, Italy

CARPINTERI, A.
Department of Structural
Engineering,
Politecnico di Torino,
10129 Torino, Italy

CARSWELL, W. S.
National Engineering Laboratory,
East Kilbride,
Glasgow G75 OQU, UK

CHOU, S. C.
US Army Laboratory Command,
Materials Technology
Laboratory,
Watertown, Massachusetts
02172-0001, USA

DVORAK, G. J.
Department of Civil Engineering,
Rensselaer Polytechnic Institute,
Troy, New York 12180, USA

179

GDOUTOS, E. E.
School of Engineering,
Democritus University of Thrace,
GR-671 00 Xanthi, Greece

JONES, R.
Structures Division,
Aero Research Laboratories,
PO Box 4331,
Melbourne, Australia

KOHN, R. V.
Courant Institute of
  Mathematical Sciences,
251 Mercer Street,
New York, New York 10012, USA

LAWS, N.
Department of Mechanical
  Engineering,
University of Pittsburgh,
Pittsburgh, Pennsylvania 15261,
  USA

MARSHALL, I. H.
Department of Mechanical
  Engineering,
Paisley College of Technology,
Paisley, Renfrewshire PA1 2BE,
  UK

OERTEL, F. H., Jr.
US Army Research Development
  and Standardization Group
  (UK), London NW1 5TH,
  UK

SIH, G. C.
Institute of Fracture and Solid
  Mechanics,
Lehigh University,
Bethlehem, Pennsylvania 18015,
  USA

SMITH, G. F.
Center for the Application of
  Mathematics,
Lehigh University,
Bethlehem, Pennsylvania 18015,
  USA

SPENCER, A. J. M.
Department of Theoretical
  Mechanics,
University of Nottingham,
Nottingham NG7 2RD, UK

TAY, T. E.
Department of Mechanical
  Engineering,
Melbourne University,
Parkville, Victoria, Australia

WILLIAMS, J.
Department of Mechanical
  Engineering,
Paisley College of Technology,
Paisley, Renfrewshire PA1 2BE,
  UK

WILLIAMS, J. F.
Department of Mechanical
  Engineering,
Melbourne University,
Parkville, Victoria, Australia

WILLIAMS, J. G.
Department of Mechanical
Engineering,
Imperial College,
London SW7 2BX, UK

WU, J. J.
US Army Research Development
and Standardization Group
(UK),
London NW1 5TH, UK

Index

Aircraft, heating effects, 56
Aluminium alloys
  temperature change in tension, 5
  thermoelastic characteristics, 54
Anisotropic laminae, analysis of,
  151–3
AS/3501–06 composites
  orthotropic elastic properties, 54
  progressive cracking of, 97
Astroidal inclusions, 32
Atomic, meaning of term, 17
Axisymmetric/Planar Elastic
      Structures (APES) finite
      element program, 119

Beltrami's criterion, 118
Bismaleimide, carbon fibre reinforced
      composite, mixed mode
      failure of, 70
Brittle fracture, dimensional
      transition from plastic
      collapse, 112, 113–15, 123
Buckingham's theorem, 121

Carbon fibre reinforced composites
  mixed mode failure of, 69–70
  see also Graphite . . .
Centre cracked slab in tension
  analysis of, 119–24
  collapse/fracture mode of, 113–14
Chopped strand mat (CSM)
      composites, residual strength
      after impact damage, 145
Classical Laminate Theory, 148

Classical thin plate theory
  bending deformations under edge
    loading, 149
  in-plane displacement components,
    152
  in-plate stretching deformations,
    149
Complex eigenvalue, FEIM used for,
    105–6
Complexity, composite behavior, 2
Concentric sphere construction, 164
Concrete
  mechanical damage model for, 116
  repeated compression loading of,
    126–7, 128
  temperature changes under cyclic
    compression loading, 127,
    128
Constitutive equations, transversely
    isotropic materials, 71, 73–9
Cooling – heating effect
  aluminium alloys, 5
  concrete, 126–7, 128
  diagnostic use of, 130
  polycarbonate, 6
  poly (methyl methacrylate), 129
  repeated loading, 53–5, 125–30
    explanation of, 130
    results, 128–30
    test procedure, 125–8
Cooling – heating effects, *see also*
    Thermomechanical
    behavio(u)r
Crack propagation, strain energy
    density theory applied to,
    118

Creep, microstructure effects, 3
Cross-ply composites, loss of stiffness
in, 95
Cross-ply laminates
axial loading of, 92
schematic of composite element, 11
Cyclic loading, thermomechanical
interactions in, 53–5, 125–30

Damage analysis, 4, 10–18
composite element, 11–12
impact damage, 135–44
Damage tolerance, meaning of term,
134
Debonding
fiber/matrix interface, 9, 10, 14
multiphase materials, 26
Degenerate Hooke's law, 163
Delamination
FEIM analysis of singularities at,
108
layered composite, 9
method of analysis for, 62–5
mixed mode tests, 65–70
one-dimensional model for, 62–3
Ductility, size effects on, 125
Duhamel – Neuman law, 51

Edge effects, exact elastic stress
analysis, 153–4
Effective moduli, 155
applications to structural
optimization, 169–73
bounds in terms of volume
fractions, 160
'density' result used, 160, 165
explicitly computable, construction
of mixtures with, 164–9
mathematically precise definitions
of, 156–60
new methods for bounding of,
160–4
physical equivalent of, 159
Eigenvalue problems, FEIM used for,
105–6

Electrochemical interface, 7
energy density contours detected by
EDI, 21, 22
Electromagnetic discharge imaging
(EDI) technique, 14
applications of, 19
damage analysis by, 14, 19
interpretation of, 19–22
technique described, 19
Energy
conservation equation, 50
dissipation, 3–4
experimental measurement of, 12
release rate approach, 55–6
Entropy, rate of change of, 51–3
Epoxy resins, carbon fibre reinforced
composite, mixed mode
failure of, 69
Exact elastic stress analysis
anisotropic laminate, 151–3
edge effects in, 153–4
isotropic laminates, 148–51

Failure
analysis
cuspidal-cornered rigid
inclusions, 27–34
partially bonded rigid inclusions,
35–44
mechanisms, 10–14
characterization of, 14–16
scale of observation of, 16–18
Fibre
breaking of, 14
time response affected by, 16
reinforced composites
hypothetical characterization of
microelements for, 15
hypothetical microfailure in, 14
thermal – mechanical interaction of,
4–5
Finite element interactive method
(FEIM), 102–4
applications of, 108
finite element mesh used, 103
global interpretation of, 106–7
identification with power method,
104–5

Finite element interactive method
(FEIM) —*contd.*
procedure used, 102–3
verification of, 107–8
Fourier analysis, 162
Fracture mechanics, 55–8
energy release rate approach, 55–6
progressive crack analysis, 96–7
strain energy density approach,
57–8

G-closure problem, 160
G-convergence approach, 156, 158
see also Homogenization
Graphite – epoxy composites
impact damage, 57
orthotropic elastic properties, 54
progressive cracking of, 97–8
see also Carbon . . .
Group representation theory
methods, 75

Hashin – Shtrikman variational
principle, 160
derivation of, 162
method of bounding based on, 161
Hilbert formulation, rigid inclusion
problem, 36–7
Homogenization, 156
concrete compared with plexiglass,
130
Hypocycloidal inclusions, 32–4

Impact damage
energy available for, 136–7
mathematical modelling of
difficulties encountered, 135
residual strength indications,
144–5
theory used, 135–44
permanent versus maximum
indentation, 139
Impact damaged laminate, analysis of,
57
Inclusions. *See* Rigid inclusions

Interface
crack considered by FEIM, 103, 107
debonding of, 9, 14
time response affected by, 16
electrochemical interface, 7
load transfer across, 8–9
mechanical interface, 7
oscillation in energy due to loading,
8–9
thermal – mechanical interaction of,
7–9
Interlaminar shear stress, 151
Intermolecular forces, 7
Irregularly shaped reinforcement, 26,
46–7
Isotropic laminates, analysis of,
148–51

Kirlian photography, 19
see also Electromagnetic discharge
imaging (EDI) technique

Layered microstructure composites,
164
Linear elastic fracture mechanics
(LEFM), 62
brittle/plastic failure transition, 123
impact damage calculations, 143
Linear inclusions, 29–31

Macro, meaning of term, 17
Macro-approach, 25
Mass conservation equation, 50
Material testing methods
inadequacy of, 3
see also Non-destructive testing
Mathematical modelling
failure mechanisms, 135
impact damage, 135–44
recent progress in, 155–73
Matrix
cracking of, 14
time response affected by, 16
thermal – mechanical interaction of,
5–7
transverse cracking in, 91–5

Matrix form, constitutive equation, 71, 73
Mechanical damage model, 115–16
Mechanical interface, 7
Micro-approach, 26
Minimum complementary energy, principle of, 172
Minimum strain energy density criterion, 28–9, 30
Mixed mode failure
  analytical methods used, 62–5
  experimental results obtained, 67–70
  test configurations used, 65–6
Moisture, thermomechanical behaviour affected by, 58
Momentum conservation equation, 50
Multiphase materials, failure modes of, 26

Non-destructive testing (NDT) techniques, 19

Optimal structures, calculation of, 169

Participants listed, 179–81
Paul's bound, 161, 166
Perforated composites, 169–72
Periodic composite, meaning of term, 156
Periodic structure, meaning of term, 157
Planar isotropic materials, impact damage modelling for, 134–45
Plastic collapse, dimensional transition to brittle fracture, 112, 113–15, 123
Plemelj function, 37
Plexiglass, temperature changes under cyclic compression loading, 129
Polycarbonate, temperature change intension, 6

Poly (ether ether ketone) (PEEK), carbon fibre reinforced composite, mixed mode failure of, 69
Poly (methyl methacrylate) (PMMA), temperature changes under cyclic compression loading, 129
Portland cement concrete, compression loading of, 126–7
Power Sweep method, FEIM compared with, 104–5
Product tables
  transverse isotropy groups, 79
  T_1, 80
  T_2, 81
  T_3, 83
  T_4, 85
  T_5, 86
Progressive transverse cracking
  analysis of, 95–8
  experimental data for graphite–epoxy composites, 97–8

Random composite, meaning of term, 156
Random structure, meaning of term, 157
Rayleigh quotient approach, 105
Relaxation techniques, 171
  advantages of, 171
  execution of, 172
Residual strength
  impact-damaged plates, 144–5
  thick plates, 144
  thin plates, 145
Rigid inclusions
  astroidal inclusions, 32
  cuspidal-cornered, 27–34
    minimum strain energy density criterion used, 28–9
    stress field of, 27–8
  hypocycloidal inclusions, 32–4
  linear inclusions, 29–31
  partially bonded, 35–44
    Hilbert formulation of, 36–7
    local stress distribution for, 40–2

partially bonded —*contd.*
   statement of problem, 35–6
   unknown coefficients determined
     for, 37–9
   square inclusions, 42–3
   triangular inclusions, 43–4

Scale of observation, failure modes
   affected by, 16–18
Semi-elliptical surface crack, impact
   damage resulting in, 143
Shear lag theory, 91, 95
Singularity fields, 102
Size
   collapse/fracture mode affected by,
     112, 113–15, 123
   ductility affected by, 125
   strength affected by, 124–5
   toughness affected by, 124–5
Spherical inclusions, 157
Spherical indentor
   impact damage caused by, 142–3
   stresses caused by, 141
   surface crack caused by, 143
Square inclusions, 42–3
Strain energy density
   factor
     crack growth affected by size, 122
     critical value of, 120
     dimensionless representation of,
       121
   theory, 57–8
     centre cracked slab in tension
       analysed using, 119–24
     cooling – heating effects
       predicted by, 5, 6, 7, 57–8
     crack growth analysed using, 118
     strain-softening behaviour and,
       115–18
Strain-softening behaviour, 115–18
Strength, size effects on, 124–5
Structural optimization, applications
   of effective moduli to,
     169–73
Substructuring method, FEIM as, 106

T300/934 laminates, progressive
   cracking of, 98

Thermomechanical behavior, 4–9,
   49–50
   basic equations for, 50–3
   constitutive equation for, 50
   fiber, 4–5
   implications for cyclic stressing,
     53–5, 125–30
   interface, 7–9
   matrix, 5–7
   see also Cooling – heating effect
Thick composite plates
   impact on
     displacement – time history for, 136
     force – time history for, 136
     residual strength for, 144
Thin composite plates
   impact on
     displacement – time history for,
       137
     force – time history for, 137
     residual strength for, 145
   loading and unloading of, 138
Time response characteristics, 15–16
Titanium, aluminium – vanadium
   alloy, thermoelastic
   characteristics, 54
Toughness, size effects on, 124–5
Translation invariance, 157, 159
Transverse cracking, 91–5
   progressive cracking, analysis of,
     95–8
Transverse isotropy groups
   definition of, 72
   group T_1, 80
     transformation of vector under,
       75–6
   group T_2, 80–2
   group T_3, 82
   group T_4, 82, 84
   group T_5, 84–5
   irreducible representations for,
     79–86
Transversely isotropic materials
   axial vector-valued function for,
     88–9
   constitutive expressions for, 73–9
     applications of, 87–90
   polar vector-valued function for,
     87–8

Transversely isotropic materials
 —*contd.*
 second-order tensor-valued
  function for, 87
Triangular inclusions, 43–4

Unidirectional fibre reinforced
 composites
 failure modes in, 10

Unidirectional fibre reinforced
 composites —*contd.*
 mixed mode failure of, 67–70

Van der Waals bonds, 7
Variation principles, 160–1

Weight minimization, 169–73